The Open University

S342

Science: a third level course

PHYSICAL CHEMISTRY

PRINCIPLES OF CHEMICAL CHANGE

BLOCK 3
REACTION MECHANISMS

THE S342 COURSE TEAM

CHAIR AND GENERAL EDITOR

Kiki Warr

AUTHORS

Keith Bolton (Block 8; Topic Study 3)

Angela Chapman (Block 4)

Eleanor Crabb (Block 5; Topic Study 2)

Charlie Harding (Block 6; Topic Study 2)

Clive McKee (Block 6)

Michael Mortimer (Blocks 2, 3 and 5)

Kiki Warr (Blocks 1, 4, 7 and 8; Topic Study 1)

Ruth Williams (Block 3)

Other authors whose previous S342 contribution has been of considerable value in the preparation of this Course

Lesley Smart (Block 6)

Peter Taylor (Blocks 3 and 4)

Dr J. M. West (University of Sheffield, Topic Study 3)

COURSE MANAGER

Mike Bullivant

EDITORS

Ian Nuttall

Dick Sharp

BBC

David Jackson

Ian Thomas

GRAPHIC DESIGN

Debbie Crouch (Designer)

Howard Taylor (Graphic Artist)

COURSE READER

Dr Clive McKee

COURSE ASSESSOR

Professor P. G. Ashmore (original course)

Dr David Whan (revised course)

SECRETARIAL SUPPORT

Debbie Gingell (Course Secretary)

Jenny Burrage

Margaret Careford

Shirley Foster

The Open University, Walton Hall, Milton Keynes, MK7 6AA.

Copyright © 1985, 1996 The Open University. First published 1985; second edition 1996. Reprinted 2002.

Edited, designed and typeset by The Open University.

Printed in the United Kingdom by Henry Ling Ltd, The Dorset Press, Dorchester DT1 1HD.

ISBN 0 7492 5179 4

This text forms part of an Open University Third Level Course. If you would like a copy of *Studying with The Open University*, please write to the Central Enquiry Service, PO Box 200, The Open University, Walton Hall, Milton Keynes, MK7 6YZ. If you have not enrolled on the Course and would like to buy this or other Open University material, please write to Open University Educational Enterprises Ltd, 12 Cofferidge Close, Stony Stratford, Milton Keynes, MK11 1BY, United Kingdom.

S342blocks3,4i2.2

CONTENTS

1 REACTION MECHANISMS:
AN INTRODUCTION

The conventional definition of the word 'mechanism' relates to the action by which a particular result is obtained, be it via a machine or a system. The mechanism by which a Bill gets through Parliament relates to the various stages it must go through from inception to its Third Reading. In chemistry, mechanism has roughly the same meaning; it describes the manner in which a reaction proceeds, from reactants to products, at the *molecular level*.

To be more specific, the term mechanism is used to describe the one or more elementary steps that determine the route between reactants and products. You met an example of this at the start of Block 2, when considering the reaction between iodide ion, I^-, and hypochlorite ion, ClO^-, in aqueous solution:

$$I^-(aq) + ClO^-(aq) = IO^-(aq) + Cl^-(aq) \tag{1}$$

Experimental evidence suggests that this reaction does *not* occur in a single elementary step, but rather by a series of such steps — the **reaction mechanism** — whose *net* result is the stoichiometric equation. A likely mechanism is:

$$ClO^- + H_2O \rightleftharpoons HClO + OH^- \tag{2}$$

$$HClO + I^- \longrightarrow HIO + Cl^- \tag{3}$$

$$OH^- + HIO \longrightarrow H_2O + IO^- \tag{4}$$

A reaction such as this, which proceeds via *more than one* elementary step, is known as a **composite reaction**, and its mechanism, equations 2, 3 and 4, is referred to as a **composite reaction mechanism**, or simply a **composite mechanism**.

Chemical reactions involve the making and breaking of bonds or the exchange of electrons. Elementary reactions, which involve only one or two, or very rarely three, reacting species, result in the making or breaking of only a *few bonds* in each step. It is hardly surprising, therefore, that a large number of chemical transformations require a series of elementary processes, even when the overall stoichiometry of the reaction is apparently very simple. Clearly, a study of composite reactions is fundamental to our understanding of chemistry.

In this Block we shall extend to composite processes the ideas developed in Block 2 for the kinetic treatment of elementary reactions. In so doing, we shall discover the key role played by kinetic studies in *establishing* the mechanism of a reaction. An understanding of mechanism provides not only chemical insight, but it may also suggest ways of making a reaction more efficient. A knowledge of reaction mechanism is also very important in *catalysis* and this is an area that is taken up in the Blocks that follow.

We begin our study of composite reactions by raising two key issues: 'How do we decide whether a reaction is composite or not?', and if it is, 'How do we establish the form of its mechanism?'.

STUDY COMMENT   Video band 3 (*Reaction mechanisms*) is associated with this Block. It is appropriate to view this sequence (which also revises some of the material in Block 2) after you have finished working through Section 5. However, you may also find it very useful to watch the sequence again when you have completed this Block.

AV sequences play a particularly important role in this Block and you should incorporate them in your study plan. The first sequence (Section 5 of the AV Booklet and band 5 on audio-cassette 2) is concerned with the kinetics of a chemical reaction that exhibits time-dependent stoichiometry. This material is **not** covered elsewhere in the Block and it is essential that you work through the sequence at the start of Section 4. The second sequence (Section 6 of the AV Booklet and band 6 on audiocassette 2) is concerned with two very important concepts that relate to composite reaction mechanisms; these are rate-limiting steps and pre-equilibria. Their discussion forms an **integral** part of Section 6.

Finally, a word about the importance of SAQs in this Block. Much of the Block concentrates on the derivation of theoretical chemical rate equations based on proposed reaction mechanisms for reactions that exhibit time-independent stoichiometry. The theoretical chemical rate equations that are derived are then compared with those found experimentally. The SAQs play an essential role in helping you to develop and practise skills in this area; for this reason you should make every effort to work through them as you progress through the Block.

2 DISCOVERING COMPOSITE REACTION MECHANISMS

In this Section we examine how the individual steps in a reaction mechanism, such as that given by equations 2, 3 and 4, are identified. It should be clear from the outset that a stoichiometric equation tells us *nothing* of the detailed reaction pathway. This point is made more forcibly by the following reaction:

$$2H_2(g) + O_2(g) = 2H_2O(g) \tag{5}$$

for which a mechanism involving up to 40 elementary steps has been suggested! To outline how such sequences are determined, we begin by considering the mechanism of a fairly simple system. From this, we develop a common strategy that can be applied to *all* reactions.

As our example we take the oxidation of nitrous acid, HNO_2, by hydrogen peroxide, H_2O_2, to give nitric acid, HNO_3, and water:

$$HNO_2(aq) + H_2O_2(aq) = HNO_3(aq) + H_2O(l) \tag{6}$$

Have a look at Table 1 and Figure 1, which show how the concentrations of the reactants and products vary during the reaction. Do you notice anything strange?

The thing that should immediately strike you is that the stoichiometry of the reaction *changes* with time. For example, after 0.5 s of reaction, 0.24 mol dm^{-3} of each reactant has been *consumed*, but only 0.01 mol dm^{-3} of nitric acid has been formed. Eventually, as indicated by the 'infinity sign' in Table 1, 0.30 mol dm^{-3} of nitric acid *is* formed, which is what you would expect from the stoichiometry in equation 6. But it seems that this equation, which indicates a 1 : 1 stoichiometry between reactants and products, applies *only* at the beginning and end of the reaction: it does not apply *throughout* the reaction. In these circumstances, the reaction is said to have **time-dependent stoichiometry**. For such a reaction it is not possible to describe its progress in terms of the reaction variable, nor is there a simple relationship between the rate of change of concentration of a reactant species with time compared with that for a product species.

Table 1 The oxidation of nitrous acid to nitric acid by hydrogen peroxide at 273.8 K[*]

time	[HNO$_2$]	[H$_2$O$_2$]	[HNO$_3$]
s	mol dm^{-3}	mol dm^{-3}	mol dm^{-3}
0	0.30	0.30	0.00
0.1	0.17	0.17	0.00
0.5	0.06	0.06	0.01
1.0	0.04	0.04	0.02
10.0	0.00	0.00	0.16
20.0	0.00	0.00	0.24
∞	0.00	0.00	0.30

[*] The concentration of the other product, water, can be assumed to be constant, since it is the solvent.

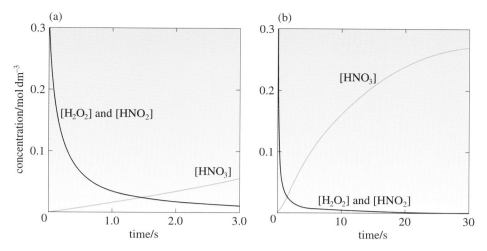

Figure 1 The variation in the concentration of reactants and products as a function of time for the oxidation of nitrous acid by hydrogen peroxide: (a) during the first 3 s of reaction, (b) during the first 30 s of reaction.

◼ Is the variation of the concentration of reactants and products with time consistent with the idea that the reaction proceeds via a single elementary step?

▨ No. The behaviour shown in Table 1 and Figure 1 cannot be rationalized in terms of a single elementary step. As you saw in Block 2, such processes have *time-independent* stoichiometry.

This leads us to our first conclusion about the mechanism of this reaction: *it does not occur in a single elementary step, and so presumably it has a composite reaction mechanism.*

Having established that we are dealing with a composite process, our next task is to establish the form of the mechanism. Let's start by reviewing what we know of this reaction. Well, it is clear from Figure 1 that the reactants quickly disappear, whereas the product, nitric acid, is formed quite slowly. As matter must be conserved, these observations suggest that the reactants first form some as yet unidentified substances (or substance), which then go(es) on to form the products.

◼ Using chemical and spectroscopic techniques, an isomer of nitric acid, pernitrous acid, HOONO, has been identified in the reaction mixture during the course of the reaction. What composite mechanism might account for this observation?

▨ The simplest mechanism that is consistent with the available evidence would be one involving a series of just two elementary steps, so that the net result is the stoichiometric equation:

$$HNO_2 + H_2O_2 \longrightarrow HOONO + H_2O \tag{7}$$

$$HOONO \longrightarrow HNO_3 \tag{8}$$

In the first step, equation 7, nitrous acid and hydrogen peroxide combine to form pernitrous acid and water. Pernitrous acid then rearranges to give nitric acid in the second step, equation 8. You may have noticed that we have left out the state symbols in equations 7 and 8. This is not forgetfulness — it is customary not to include 'states' when writing out a detailed mechanism.

Compounds like pernitrous acid, which are formed in one step and consumed in another, are known as **intermediates**. Figure 2 shows how the concentration of pernitrous acid varies with time during the reaction. Such behaviour is fairly characteristic of intermediates, the shape of the curve reflecting the *temporary* nature of such species. However, it should be emphasized that an intermediate must *not* be confused with an activated complex. The latter occurs at the top of the energy barrier for an *elementary* reaction and has an entirely transitory existence. Intermediates can have quite long lifetimes, as shown by the build-up of pernitrous acid during the reaction indicated in Figure 2.

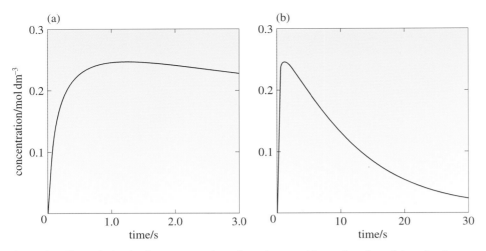

Figure 2 The variation in the concentration of pernitrous acid as a function of time for the oxidation of nitrous acid by hydrogen peroxide: (a) during the first 3 s of reaction, (b) during the first 30 s of reaction.

We shall see later that the form of the plots shown in Figures 1 and 2 can be interpreted by considering the relative rates of the two steps (equations 7 and 8) in the mechanism. But, before we go on to analyse the kinetics of this and other reactions, we shall spend a little more time examining the approach we took in establishing a mechanism for reaction 6. From this, we shall develop a common strategy for deciding whether or not a reaction is composite and then, if it is composite, going on to discover the form of the mechanism.

2.1 A strategy for establishing a composite reaction mechanism

Although the detailed procedure for determining a composite mechanism will vary from case to case, the overall approach can be broken down into three stages:

- finding out if the reaction is composite;

- proposing a mechanism for it;

- confirming the proposed mechanism.

It is useful to review each of these in turn; the strategy is summarized in Box 1 (p. 14).

2.1.1 The first stage: is the reaction composite?

As you have just seen, the first stage in dealing with any reaction is to establish if it is indeed composite. This is usually achieved by demonstrating that the reaction is *not elementary*, all reactions being assumed to be elementary until shown to be otherwise. In our example based on reaction 6, the variation of the concentrations of the reactants and products with time clearly indicated that the reaction could not occur in a single step. Unfortunately, few reactions give up their secrets so easily! As a consequence, many composite reactions can go unnoticed for years. You have already met one example of this in Block 2: for sixty years the gas-phase reaction between hydrogen and iodine was regarded as the textbook example of an elementary reaction. However, after careful re-examination it was eventually shown to occur via a series of steps.

To establish whether a reaction is elementary or not, chemists have to use a range of techniques from their armoury. Below we outline, albeit briefly, just two methods that we shall employ in this Block to reveal a composite mechanism.

Detection of intermediates

Any reaction that occurs via a series of steps will involve intermediate species. It ought, therefore, to be possible to establish that a reaction is not elementary simply by identifying intermediate(s). Both spectroscopic and chemical techniques can be used to characterize such species. However, this approach is often limited by the fact that many intermediates are extremely reactive and short lived, which makes them difficult to identify in the reaction mixture. In such cases the concentration of an intermediate can be so low that, within the accuracy of the chemical analysis, the reaction effectively exhibits time-independent stoichiometry. We shall meet examples like this in later Sections.

Kinetic data

The most commonly used method of identifying a composite reaction is an examination of the experimental rate equation. In Block 2 it was stated that *for all elementary reactions that you will encounter in this Course* it will be safe to assume that the chemical rate equation can be written down by inspecting the stoichiometry.

■ If the following reactions are elementary, what form would you expect the chemical rate equation to take?

(a) $CH_3NC(g) \longrightarrow CH_3CN(g)$ (9)

(b) $H_2(g) + F\cdot(g) \longrightarrow HF(g) + H\cdot(g)$ (10)

(c) $2Br\cdot(g) \longrightarrow Br_2(g)$ (11)

■ (a) $J = k_R[CH_3NC]$ (12)

(b) $J = k_R[H_2][F\cdot]$ (13)

(c) $J = k_R[Br\cdot]^2$ (14)

The corollary of the above statement is that if an experimental rate equation differs from that 'expected' from the stoichiometry (assuming that the reaction were elementary), then the reaction must proceed via a composite mechanism. Try out the following examples to see if you can determine which reactions are composite.

■ Under certain conditions, the reaction between 2-chloro-2-methylpropane, $(CH_3)_3CCl$, and hydroxide ion, OH^-,

$(CH_3)_3CCl(aq) + OH^-(aq) = (CH_3)_3COH(aq) + Cl^-(aq)$ (15)

has the following experimental rate equation:

$J = k_R[(CH_3)_3CCl]$ (16)

Is this a composite reaction?

■ If this reaction were elementary, the stoichiometry would indicate a chemical rate equation of the form:

$J = k_R[(CH_3)_3CCl][OH^-]$ (17)

Since the experimental rate equation does *not* involve the concentration of hydroxide ion, $[OH^-]$, this points to a composite reaction.

■ The conversion of hypochlorite ion, ClO^-, to chlorate ion, ClO_3^-,

$3ClO^-(aq) = ClO_3^-(aq) + 2Cl^-(aq)$ (18)

has the experimental rate equation:

$J = k_R[ClO^-]^2$ (19)

Is this a composite reaction?

■ If this reaction were elementary, the chemical rate equation expected from the stoichiometry would have the form:

$J = k_R[ClO^-]^3$ (20)

This is different from that obtained experimentally, which suggests that this reaction is composite.

But also notice in this example that even if the experimental rate equation were similar to equation 20, it is doubtful whether such a reaction could be elementary, because, as you saw in Block 2, collisions involving three molecules are very rare indeed. This leads to the general rule that *any reaction having a stoichiometric equation involving more than three reactant species cannot possibly occur via a single elementary process, and thus must involve a composite mechanism.*

■   The reaction between carbon monoxide, CO, and chlorine, Cl_2,

$$CO(g) + Cl_2(g) = COCl_2(g) \qquad (21)$$

has the experimental rate equation:

$$J = k_R[CO][Cl_2]^{3/2} \qquad (22)$$

Is this a composite reaction?

■   The chemical rate equation expected from the stoichiometry is:

$$J = k_R[CO][Cl_2] \qquad (23)$$

This is at odds with that obtained experimentally, thus indicating that the reaction involves a composite mechanism.

This example highlights another important point. Since the partial order of a reactant in an elementary reaction is always positive and integral, *any reaction with an experimental rate equation involving fractional or negative partial orders must be a composite reaction.*

STUDY COMMENT   We have now completed the first stage of the strategy for establishing a composite reaction mechanism. For convenience, we have summarized the criteria for concluding that a reaction is composite in the box below. ▬▬▬▬▬

CRITERIA FOR CONCLUDING THAT A REACTION IS COMPOSITE

If a reaction manifests one or more of the following features, then it is deemed to be composite:

1   the detection of an intermediate;

2   a stoichiometric equation involving more than three reactant species;

3   an experimental equation that differs from that expected from the stoichiometry of the reaction, for example one involving partial or negative orders.

2.1.2   The second stage: proposing a composite mechanism

Once a reaction has been demonstrated *not* to be elementary, the next stage is to propose a likely composite reaction mechanism that will account for all the facts. We begin this Section by reviewing the *general* types of composite mechanism that you will meet in this Course. We then go on to discuss how to choose the most likely candidate.

The mechanism that was suggested earlier for the oxidation of nitrous acid by hydrogen peroxide, equations 7 and 8, can be rewritten in an alphabetical form:

$$A + B \longrightarrow C + D \qquad (24)$$

$$C \longrightarrow E \qquad (25)$$

This general type of mechanism is known as a two-step **consecutive mechanism**. One shorthand method of writing equations 24 and 25 is:

$$A + B \longrightarrow \begin{matrix} C \longrightarrow E \\ + \\ D \end{matrix} \qquad (26)$$

However, this can sometimes lead to confusion and so, for the time being, we shall present mechanisms as individual steps. (As a matter of information, it is worth noting that in this and the next Block we shall mainly limit our discussion to consecutive mechanisms involving just two or three steps.)

■ What do the arrows ($\longrightarrow$) in equations 24 and 25 imply?

▨ The arrows imply that the reactions are thought to be elementary.

Sometimes, both the forward and reverse elementary steps of a reaction are important, and in this case two arrows are used:

$$A + B \underset{\longleftarrow}{\longrightarrow} C + D \qquad\qquad (27)$$

$$C \longrightarrow E \qquad\qquad (28)$$

Elementary reactions occurring in the forward and reverse direction, as in equation 27, are described as **opposing reactions**.

When a reactant can undergo two separate elementary steps to give two products, as shown below, the reactions are described as **parallel reactions**:

$$A \longrightarrow B \qquad\qquad (29)$$

$$A \longrightarrow C \qquad\qquad (30)$$

The final type of composite mechanism that you will meet in this Course is a *chain reaction*; specific examples will be dealt with in Section 7.

Having described the various types of mechanism, we now move on to discuss the method of choosing the *most likely* candidate. For the oxidation of nitrous acid, this task was straightforward. However, as you will appreciate, it can be quite daunting with more complex systems. The process is essentially one of trial and error. But there are some guidelines: in fact, it is possible to put forward four criteria that can be used to weed out unsuitable mechanisms.

1   Firstly, the mechanism must be able to *account for all of the data that are available for the reaction*. If a number of intermediates are detected, they should all appear somewhere in the proposed scheme. This was the reasoning behind the mechanism proposed in equations 7 and 8.

Similarly, if the experimental rate equation has a particular form, then the mechanism should predict a chemical rate equation that is identical. *This criterion is a particularly powerful method for weeding out unsuitable mechanisms*. Indeed the focus of this Block is the relationship between reaction mechanisms and chemical rate equations.

2   Secondly, the mechanism must be *reasonable in terms of the chemistry that it describes*. Don't worry. This does not mean that you will be expected to know the chemistry of every reaction in this Course! When such information is required, the relevant chemistry will be discussed in detail. Each step should, of course, be balanced and not involve chemically unrealistic reactants or products, such as compounds in which the elements have unlikely oxidation numbers.

■ As each of the steps in a mechanism is elementary, what are their possible molecularities?

▨ Each step should be either *unimolecular, bimolecular* or, very occasionally, *termolecular*.

3   Thirdly, for non-chain reactions, *the sum of the various steps should add up to the overall stoichiometric equation*. Consider the reaction between iodide ion and hypochlorite ion that we discussed earlier. The stoichiometric equation is:

$$I^-(aq) + ClO^-(aq) = IO^-(aq) + Cl^-(aq) \qquad\qquad (1)$$

and the proposed mechanism is:

$$ClO^- + H_2O \underset{\longleftarrow}{\longrightarrow} HClO + OH^- \qquad\qquad (2)$$

$$HClO + I^- \longrightarrow HIO + Cl^- \qquad\qquad (3)$$

$$OH^- + HIO \longrightarrow H_2O + IO^- \qquad\qquad (4)$$

■ Is the composite mechanism consistent with the stoichiometric equation?

▨ Adding together equations 2, 3 and 4 gives

$$ClO^- + H_2O + HClO + I^- + OH^- + HIO \longrightarrow HClO + OH^- + HIO + Cl^- + H_2O + IO^-$$

Cancelling out the species that appear on both sides of the equation (that is, H_2O, $HClO$, HIO and OH^-) gives:

$$I^- + ClO^- = IO^- + Cl^- \tag{31}$$

We conclude that the mechanism *is* in agreement with the stoichiometric equation.

Notice that this type of summation only holds if *opposing reactions are considered as a single step linked by forward and reverse arrows* — as in equation 2, for example. If equation 2 is split into two separate steps:

forward: $\quad ClO^- + H_2O \longrightarrow HClO + OH^- \tag{32}$

reverse: $\quad HClO + OH^- \longrightarrow ClO^- + H_2O \tag{33}$

then summation of equations 32 and 33 does *not* give the overall stoichiometric equation.

It is also worth pointing out that very occasionally it is necessary to include one step more than once. Consider, for example, the hypothetical reaction

$$A + B = 2C \tag{34}$$

which proceeds via the mechanism

$$A + B \longrightarrow 2D \tag{35}$$

$$D \longrightarrow C \tag{36}$$

The first step produces two molecules of D, whereas the second step only describes the consumption of one molecule of D. Thus, the second step must occur *twice* for each occurrence of the first step. In this case the sum of the first step and twice that of the second step gives the stoichiometric equation, as you should be able to confirm for yourself.

4 Finally, despite the three criteria outlined already, it can still remain the case that a number of possible mechanisms are equally valid. When this happens, we should take our cue from the worthy William of Occam*, and use the simplest explanation. For example, as mentioned in Section 2.1.1, in the absence of evidence to the contrary, all reactions are assumed to be elementary; you can't get simpler than that!

STUDY COMMENT   We have now completed the second stage of the strategy for establishing a composite mechanism. The criteria associated with this second stage — eliminating unsuitable mechanisms — are summarized in the Box below.

CRITERIA FOR ELIMINATING UNSUITABLE COMPOSITE MECHANISMS

A proposed composite mechanism would be unsuitable if:

1   it does not account for all the available data or does not predict a chemical rate equation identical to the experimental rate equation;

2   it involves unrealistic chemistry;

3   the sum of the steps (for non-chain reactions) does not add up to the overall stoichiometric equation;

4   the mechanism is more complicated than an alternative mechanism.

* *Occam's razor*: a maxim attributed to William (of) Occam (*ca.* 1290–1350) — 'it is vain to do with more what can be done with fewer'; that is, if the facts resulting from an experiment can be explained without making additional hypotheses, there are no grounds for making those hypotheses.

2.1.3 The third stage: confirmation

A mechanism is best described as a model of how a reaction is *thought* to proceed. As the last Section illustrated, we should choose the simplest and most reasonable reaction mechanism that fits all the facts. Since we cannot usually monitor the pathway taken by individual molecules, our knowledge of reaction mechanism is based on *indirect* evidence. In general, the experiments that can be carried out are able to *disprove* a particular composite mechanism, but, despite having a wide variety of evidence indicating its probable validity, we can *never* be perfectly confident that it is correct. Of course, the more experiments that are in agreement with the proposed mechanism, the more *plausible* it will be. The third stage in the establishment of a mechanism is therefore to test this plausibility: it involves designing new experiments with the sole purpose of checking the proposed mechanism. If the new evidence is consistent with the mechanism, it remains unscathed. But if there are inconsistencies, it must be modified or abandoned; that is, a new mechanism that is compatible with all the known facts must be devised to take its place. The exact nature of these experiments will depend on the system under study. It is impracticable to review all the possibilities here, and we shall give just one example of a typical approach.

In Section 2.1.1, it was shown that the conversion of hypochlorite ion, ClO^-, to chlorate ion, ClO_3^-,

$$3ClO^-(aq) = ClO_3^-(aq) + 2Cl^-(aq) \tag{18}$$

is not elementary. The mechanism that has been proposed to explain the experimental rate equation is:

$$2ClO^- \longrightarrow ClO_2^- + Cl^- \tag{37}$$

$$ClO_2^- + ClO^- \longrightarrow ClO_3^- + Cl^- \tag{38}$$

One method of testing this proposal is to synthesise the intermediate chlorite ion, ClO_2^-, by an *independent route*. If this did *not* react with the hypochlorite ion in a manner consistent with the second step, this would invalidate the mechanism. In fact, chlorite ion does react with hypochlorite ion as predicted by equation 38, and thus the mechanism is not contradicted.

STUDY COMMENT   We have now completed the third and final stage of the strategy for establishing a composite reaction mechanism. A summary of this final stage — confirming a proposed mechanism — is given in the box below.

HOW TO CONFIRM A PROPOSED MECHANISM

Design experiments that will test the mechanism; for example, synthesise the suggested intermediate by an independent method and then investigate whether it reacts according to the expectations of the suggested mechanism.

2.2 Summary of Section 2

1   The oxidation of nitrous acid to nitric acid by hydrogen peroxide probably occurs via a two-step process involving pernitrous acid as the intermediate. The reaction is said to have time-dependent stoichiometry. The stoichiometry of such a reaction changes with time.

2   A common strategy for determining a composite reaction mechanism is summarized in Box 1 (overleaf).

BOX 1   Strategy for determining the mechanism of a composite reaction

Stage 1: Is the reaction composite?   This is usually achieved by proving that the reaction is not elementary. A wide variety of techniques can be employed, the principal ones being:

- *Detection of intermediates*   If an intermediate can be detected, the reaction must involve a series of steps. (Intermediates often manifest themselves through time-dependent stoichiometry.)

- *Kinetic data*   If the experimental rate equation of the reaction is not consistent with that obtained assuming the reaction is elementary (unimolecular, bimolecular or, rarely, termolecular), then a composite mechanism is involved.

Stage 2: Proposing a mechanism   The mechanisms you will meet in this Course involve consecutive, opposing, parallel and chain reactions. Choosing a mechanism is essentially a trial-and-error operation: however, the following criteria can be used to weed out unsuitable proposals:

- The mechanism must accommodate all of the data that are available for a reaction, and should predict a chemical rate equation identical to the experimental rate equation.

- The mechanism should be reasonable in terms of the chemistry it describes. Each step should be unimolecular, bimolecular or, occasionally, termolecular, and be chemically realistic.

- For non-chain reactions, the sum of the various steps must add up to the stoichiometric equation.

- If there is more than one suitable mechanism, the *simplest* is chosen.

Stage 3: Confirmation   We can never have total confidence in any mechanism, and so it is necessary to design new experiments to test the plausibility of a proposed composite mechanism.

STUDY COMMENT   The strategy summarized in Box 1 will be used several times throughout this Block. It is important that you are able to apply this strategy when given information about a particular reaction. SAQs 1 and 2 provide you with practice in applying Stages 1 and 2 in Box 1.

SAQ 1   Which of the following reactions are likely to be composite according to the experimental rate equations given?

(a) $N_2O_5(g) + NO(g) = 3NO_2(g)$ (39)

$J = k_R[N_2O_5]$ (40)

(b) $I_2(g) + H_2(g) = 2HI(g)$ (41)

$J = k_R[I_2][H_2]$ (42)

SAQ 2   Vanadium(II) and vanadium(IV) react in acidic solutions to give vanadium(III) according to equation 43:

$V^{2+}(aq) + VO^{2+}(aq) + 2H^+(aq) = 2V^{3+}(aq) + H_2O(l)$ (43)

At the beginning of the reaction, the concentration of $V^{2+}(aq)$ (reactant) decreases more rapidly than the concentration of $V^{3+}(aq)$ (product) increases.

(a)   What evidence points to a composite reaction?

(b)   If this reaction does involve a composite reaction mechanism, which of the following three mechanisms is the *most* plausible? (You may assume that the chemistry in all three mechanisms is reasonable.)

(i) $V^{2+} + VO^{2+} \longrightarrow V_2O^{4+}$ (44)

$V_2O^{4+} + H^+ \longrightarrow VOH^{2+} + V^{3+}$ (45)

$VOH^{2+} + H^+ \longrightarrow VO^{3+} + H_2$ (46)

(ii) $V^{2+} + VO^{2+} \longrightarrow V_2O^{4+}$ (47)

$V_2O^{4+} + H^+ \longrightarrow VOH^{2+} + V^{3+}$ (48)

$VOH^{2+} + H^+ \longrightarrow V^{3+} + H_2O$ (49)

(iii) $V^{2+} + VO^{2+} \longrightarrow V_2O^{4+}$ (50)

$V_2O^{4+} \longrightarrow VO^{3+} + V^+$ (51)

$V^+ + VO^{3+} + 2H^+ \longrightarrow 2V^{3+} + H_2O$ (52)

3 THE KINETICS OF COMPOSITE REACTIONS THAT EXHIBIT TIME-INDEPENDENT STOICHIOMETRY

One of the most interesting aspects of the study of reaction rates is the insight it provides into the mechanism of a reaction. In the last Section we discussed a common strategy for establishing a composite mechanism. One criterion mentioned under Stage 2 for testing a proposed mechanism was that it should *accommodate all of the data that are available for a reaction*. This includes examining whether a composite mechanism can account for the experimental rate equation or, alternatively, the variation in the concentration of reactants, intermediates and products with time. Indeed, this turns out to be the most important method of eliminating unsuitable mechanisms, and so we shall spend the rest of this Block discussing the kinetics of composite reactions.

Before we go on to examine the plausibility of a possible mechanism in a specific example, it is useful to say a little more about the notation that is used in the elementary steps of mechanisms. Consider the following hypothetical mechanism:

$$A + B \xrightarrow{k_1} C \tag{53}$$

$$C \underset{k_{-2}}{\overset{k_2}{\rightleftharpoons}} D \tag{54}$$

$$D + E \xrightarrow{k_3} F \tag{55}$$

As written, the mechanism consists of three consecutive elementary steps, the second of which is reversible. Each step has a rate constant associated with it. For a given step the rate constant is written above the arrow in the equation and, furthermore, it is given a subscript. This subscript provides a convenient, as well as consistent, means of identifying the rate constant, since it indicates the position of the step in the overall sequence; that is, we use a subscript 1 for the first step, a subscript 2 for the second step, and so on. In the second reversible step the subscript is *positive* for reaction going in the *forward* direction, and *negative* for the rate constant associated with the *reverse* reaction. We shall use this notation throughout the Course.

We begin by looking in detail at one of the examples in SAQ 1 — the reaction of dinitrogen pentoxide, N_2O_5, with nitrogen monoxide (nitric oxide), NO, in the gas phase:

$$N_2O_5(g) + NO(g) = 3NO_2(g) \tag{39}$$

This reaction exhibits *time-independent stoichiometry* and has an experimentally determined rate equation of the form:

$$J = -\frac{d[N_2O_5]}{dt} = -\frac{d[NO]}{dt} = \frac{1}{3}\frac{d[NO_2]}{dt} = k_R[N_2O_5] \qquad (56)$$

Having established that this reaction is not an elementary process, the next stage (using the strategy given in Box 1) is to propose a plausible mechanism; one possible candidate is:

$$N_2O_5 \xrightarrow{\ k_1\ } NO_2 + NO_3 \qquad (57)$$

$$NO + NO_3 \xrightarrow{\ k_2\ } 2NO_2 \qquad (58)$$

■ Is there an intermediate in this mechanism? If so what is it?

▨ The intermediate is NO_3, which is formed in the first step and consumed in the second.

■ Applying the criteria given in Stage 2 in Box 1, is this mechanism plausible? (You may assume that the 'chemistry' involved in the two steps is reasonable.)

▨ Yes, as far as we can say. The sum of the two steps gives the stoichiometric equation and each of the steps is either unimolecular or bimolecular. As yet, we cannot say whether this mechanism can account for the experimental rate equation given by equation 56.

In the rest of Section 3 we examine how to derive the **chemical rate equation predicted by a mechanism**. This describes how the rate of reaction, J, would depend on the concentrations of the various species, if the proposed mechanism were valid. Comparison of this theoretical chemical rate equation with the experimental rate equation will show whether or not a mechanism is plausible. We shall continue to use the reaction between N_2O_5 and NO in the gas phase (equation 39) as our main example.

3.1 Theoretical rate equations for composite reactions

The first stage in any kinetic analysis of a composite mechanism is to write down expressions for the rate of change in concentration of reactants, products and intermediate species, based on an examination of how the concentration of each of these is affected by the elementary steps of the mechanism. This may sound very complicated, but it is in fact quite a simple task. Take, for example, the reactant dinitrogen pentoxide, N_2O_5, in equation 39. This is consumed in the first step of the proposed mechanism (equation 57), but its concentration is *not* changed as a result of the second step (equation 58). Thus, the expression for the overall rate of change in concentration of N_2O_5 predicted by this mechanism is the *same* as that predicted by the first elementary step alone, and can be written down from the stoichiometry and the definition of J for this step:

$$-\frac{d[N_2O_5]}{dt} = k_1[N_2O_5] \qquad (59)$$

Equation 59 is referred to as the **theoretical rate equation** *for dinitrogen pentoxide*. You may notice that this equation is identical to the experimentally determined rate equation 56, with $k_1 = k_R$. This is an observation we shall return to later. But first, what about the theoretical rate equation for nitric oxide (nitrogen monoxide)?

■ How is the concentration of NO affected by the two steps in the mechanism?

▨ There is no change in the concentration of NO as a result of the first step (equation 57). However, NO is consumed in the second step (equation 58).

■ What is the theoretical rate equation for NO as predicted by this mechanism?

The theoretical rate equation will be the same as that predicted by the second elementary step; that is

$$-\frac{d[NO]}{dt} = k_2[NO][NO_3] \tag{60}$$

This theoretical rate equation does not bear much resemblance to the experimental rate equation (equation 56). However, before we set about further analysis of why this should be, let's first of all go ahead and consider the theoretical rate equations for NO_2 and NO_3.

The product, NO_2, is formed in both the first step and the second step, so to determine the overall variation of its concentration with time we must consider how it is affected by each step. Treating the first step (equation 57) in *isolation*, we can write

$$\left(\frac{d[NO_2]}{dt}\right)_1 = k_1[N_2O_5] \tag{61}$$

The term $(d[NO_2]/dt)_1$ represents the rate of change of concentration of NO_2 due to the *first step alone*. The subscript outside the bracket indicates which step is under consideration. Similarly, treating the second step (equation 58) in isolation,

$$\frac{1}{2}\left(\frac{d[NO_2]}{dt}\right)_2 = k_2[NO][NO_3] \tag{62}$$

■ Why is there a factor of $\frac{1}{2}$ in equation 62?

▨ The factor of $\frac{1}{2}$ arises from the definition of J for the second elementary step, since the stoichiometric number of NO_2 in this step is +2.

The overall rate of concentration change for NO_2 is obtained by summing the rate of concentration change due to each step:

$$\frac{d[NO_2]}{dt} = \left(\frac{d[NO_2]}{dt}\right)_1 + \left(\frac{d[NO_2]}{dt}\right)_2 \tag{63}$$

Since from equation 62,

$$\left(\frac{d[NO_2]}{dt}\right)_2 = 2k_2[NO][NO_3] \tag{64}$$

combination of equations 61, 63 and 64 gives

$$\frac{d[NO_2]}{dt} = k_1[N_2O_5] + 2k_2[NO][NO_3] \tag{65}$$

Note that the factor of 2, which arises from our definition of J for the second step, takes into account that two molecules of NO_2 are produced in the second step, whereas only one is produced in the first.

■ How is the concentration of the intermediate NO_3 affected by the two steps of the mechanism?

▨ NO_3 is formed in the first step and consumed in the second.

■ Treating each step in isolation, write down the rate equations for $(d[NO_3]/dt)_1$ and $(d[NO_3]/dt)_2$.

▨
$$\left(\frac{d[NO_3]}{dt}\right)_1 = k_1[N_2O_5] \tag{66}$$

$$-\left(\frac{d[NO_3]}{dt}\right)_2 = k_2[NO][NO_3] \tag{67}$$

Note carefully that the negative sign in equation 67 arises because NO_3 is a reactant in the second step (equation 58) and so its stoichiometric number is −1.

The overall rate of change of concentration of NO_3 is obtained by summing the rate of concentration change due to each step:

$$\frac{d[NO_3]}{dt} = \left(\frac{d[NO_3]}{dt}\right)_1 + \left(\frac{d[NO_3]}{dt}\right)_2 \qquad (68)$$

Thus, combining equations 66, 67 and 68 gives

$$\frac{d[NO_3]}{dt} = k_1[N_2O_5] - k_2[NO][NO_3] \qquad (69)$$

For convenience, the theoretical rate equations we have determined for N_2O_5, NO, NO_2 and NO_3 are collected together in Box 2.

BOX 2

$$-\frac{d[N_2O_5]}{dt} = k_1[N_2O_5] \qquad (59)$$

$$-\frac{d[NO]}{dt} = k_2[NO][NO_3] \qquad (60)$$

$$\frac{d[NO_2]}{dt} = k_1[N_2O_5] + 2k_2[NO][NO_3] \qquad (65)$$

$$\frac{d[NO_3]}{dt} = k_1[N_2O_5] - k_2[NO][NO_3] \qquad (69)$$

At present, apart from equation 59, these theoretical rate equations *bear little resemblance* to the experimental rate equation (56), and as they stand they cannot be manipulated to give such a simple solution:

$$J = -\frac{d[N_2O_5]}{dt} = -\frac{d[NO]}{dt} = \frac{1}{3}\frac{d[NO_2]}{dt} = k_R[N_2O_5] \qquad (56)$$

However, it is very important to remember that the overall reaction (equation 39) exhibits time-independent stoichiometry. This fact is crucial in enabling us to make an approximation that allows us to simplify the kinetic expressions in Box 2. The approximation is known as the *steady-state approximation*.

3.2 The steady-state approximation

One of the reasons why we cannot simplify equations 60, 65 and 69 in Box 2 is that we do not know the concentration of the intermediate NO_3 at any given time during the reaction.

- ■ Since the reaction exhibits time-independent stoichiometry, what can we say, *in qualitative terms*, about the concentration of NO_3 during the reaction?

- ■ If the reaction exhibits time-independent stoichiometry, then — within the accuracy of the chemical analysis — the relative concentrations of the reactants and products must approximate to those predicted from the stoichiometric equation. In turn, this must mean that the concentration of the intermediate, NO_3, remains relatively small throughout the course of the reaction.

In fact, the concentration of NO_3 remains relatively small throughout the course of reaction because it reacts almost as soon as it is formed. As such, it is referred to as a **reactive intermediate**.

In such circumstances, where the concentration of an intermediate never becomes significant compared with that of the reactants or products in a reaction, *we can make the assumption that the rate of change of concentration of this intermediate with time is negligible and can be approximated to zero.* This assumption is known as the **steady-state approximation**. Thus, for our particular example we can write

$$\frac{d[NO_3]}{dt} = 0 \tag{70}$$

and equation 69 therefore becomes:

$$0 = k_1[N_2O_5] - k_2[NO][NO_3] \tag{71}$$

then $k_1[N_2O_5] = k_2[NO][NO_3]$

Obviously, the steady-state approximation cannot be entirely valid in our example: at the start of the reaction the concentration of NO_3 will be zero, and (in a similar fashion to pernitrous acid, the intermediate that we met in Section 2) its concentration must build up to a maximum value and then decrease to zero again.

■ Table 2 lists calculated* values of the terms in equation 69 at different times during the reaction. From inspection of these data, what do you think is the principal justification for the steady-state approximation?

Table 2 Calculated values of the terms in equation 69 at different times during reaction 39

time	$k_1[N_2O_5]$	$k_2[NO][NO_3]$	$d[NO_3]/dt$
s	$\mathrm{mol\ dm^{-3}\ s^{-1}}$	$\mathrm{mol\ dm^{-3}\ s^{-1}}$	$\mathrm{mol\ dm^{-3}\ s^{-1}}$
0	3.010×10^{-4}	0	3.010×10^{-4}
1×10^{-8}	3.010×10^{-4}	2.103×10^{-4}	0.907×10^{-4}
1	2.228×10^{-4}	2.228×10^{-4}	5.587×10^{-13}
2	1.649×10^{-4}	1.649×10^{-4}	4.135×10^{-13}
4	0.903×10^{-4}	0.903×10^{-4}	2.265×10^{-13}
8	0.271×10^{-4}	0.271×10^{-4}	0.679×10^{-13}

■ For most of the reaction, the two terms $k_1[N_2O_5]$ (which represents the rate of formation of NO_3) and $k_2[NO][NO_3]$ (which relates to its consumption) are *almost identical, and both are considerably larger than* $d[NO_3]/dt$: thus, the error introduced by using equation 71 will be negligible.

Putting this in another, more general, way, whenever we make the assumption that d[intermediate]/dt = 0, we are *not* saying that the concentration of an intermediate does not change during a reaction, but we are saying that its rate of change is small when compared with other terms in the appropriate theoretical rate equation.

Now let's return to our example. Since the two terms $k_1[N_2O_5]$ and $k_2[NO][NO_3]$ are of similar size, we can say that for most of the reaction a 'steady state' is set up in which the rate of formation of the intermediate in the first step ($k_1[N_2O_5]$) is equal to the rate at which it is consumed in the second ($k_2[NO][NO_3]$). If this did not happen, the amount of the intermediate would build up to detectable quantities. Looking at the data in Table 2, the only time when this assumption is questionable is at the very beginning of this reaction (see for example, the second entry in Table 2); clearly the steady state is not established immediately, but takes a little time to establish itself.

Once we have made the steady-state approximation, equations such as 71 can often be rearranged such that the *concentrations of the intermediates can be expressed in terms of the concentrations of just the reactants and products* — and these, of course, are measurable quantities. Having achieved this, we can then substitute for these concentration terms in the remaining theoretical rate equations (such as 60 and 65), and thus greatly simplify them.

---

* As described in Section 5 of the AV Booklet and the accompanying tape sequence, it is possible to integrate the set of differential equations 59, 60, 65 and 69 to give a set of integrated rate equations. Using independently determined values of k_1 and k_2, we can then calculate the concentration of each species at different times during the reaction.

3.3 Simplifying rate equations with the help of the steady-state approximation

Returning to the matter in hand, we can rearrange equation 71 to give an expression for $[NO_3]$, the concentration of the intermediate:

$$[NO_3] = \frac{k_1[N_2O_5]}{k_2[NO]} \qquad (72)$$

This can then be used to simplify the remaining theoretical rate equations. For example, substituting this expression for $[NO_3]$ in equation 60 gives:

$$\frac{d[NO]}{dt} = -k_2[NO] \times \frac{k_1[N_2O_5]}{k_2[NO]} \qquad (73)$$

Cancelling terms, this becomes

$$\frac{d[NO]}{dt} = -k_1[N_2O_5] \qquad (74)$$

■ What does equation 65 reduce to on substituting for $[NO_3]$ from equation 72?

□ On substitution, equation 65 becomes:

$$\frac{d[NO_2]}{dt} = k_1[N_2O_5] + 2k_2[NO] \times \frac{k_1[N_2O_5]}{k_2[NO]} \qquad (75)$$

$$= k_1[N_2O_5] + 2k_1[N_2O_5] = 3k_1[N_2O_5] \qquad (76)$$

Rearrangement gives

$$\frac{1}{3}\frac{d[NO_2]}{dt} = k_1[N_2O_5] \qquad (77)$$

Combining equations 59, 74 and 77, gives the chemical rate equation predicted by this mechanism:

$$J = -\frac{d[N_2O_5]}{dt} = -\frac{d[NO]}{dt} = \frac{1}{3}\frac{d[NO_2]}{dt} = k_1[N_2O_5] \qquad (78)$$

■ Is equation 78 compatible with the experimental rate equation 56?

□ Yes, equations 56 and 78 are both first order in N_2O_5.

In fact, equations 56 and 78 are identical if $k_R = k_1$. This allows us to interpret the experimental rate constant in terms of the rate constant for the first elementary step. Notice that k_2 does not appear in any of these rate expressions. (We shall have more to say about this in Section 6 of the AV Booklet and the accompanying tape sequence.)

To summarize: we set out with the aim of using kinetic data to test the two-step mechanism proposed for the gas-phase reaction between dinitrogen pentoxide, N_2O_5, and nitric oxide (nitrogen monoxide), NO (equation 39). First we wrote down the theoretical rate equations for each of the species based on the elementary steps of the proposed mechanism. Since the overall reaction exhibited time-independent stoichiometry, we were able to use the steady-state approximation to simplify these rate equations and derive a chemical rate equation consistent with that observed experimentally. Thus, our mechanism fits the facts. Although we cannot be absolutely confident that this is the correct pathway, it is certainly a very plausible one and, currently, it is accepted as the most likely mechanism for this reaction.

You may have noticed that in our detailed analysis we were able to show that the chemical rate equation (equation 78) could be derived from any *one* of the theoretical rate equations 59, 60 and 65 in Box 2. In other words, whether we select a theoretical rate equation based on a reactant species or a product species, we can still simplify it to give the same chemical rate equation. This underlines a very important point: provided we know that a reaction has time-independent stoichiometry, then it follows that the steady-state approximation applies to any intermediates formed. Under these circumstances, it is *not* necessary to write down the theoretical rate equation for *all* of the reactant and product species in order to find the chemical rate equation: a judicious selection will suffice. But how is this selection made?

The answer is that there are no hard and fast rules, but it often turns out that some expressions are much easier to simplify than others. We can see an example of this if we look back over the analysis in this Section. If we compare the theoretical rate equations 59, 60 and 65 in Box 2, which of these is the most straightforward to derive and then simplify? Consider each of them in turn:

- *Equation 59* This is easily written down, since the reactant, N_2O_5, is only involved in a single step of the mechanism. Furthermore, because the equation does not contain a term for the intermediate, no simplification is necessary!

- *Equation 60* Here the reactant NO is only involved in the second step of the mechanism. However, as the equation involves a term for the intermediate, it needs further simplification.   *which we can't measure directly!*

- *Equation 65* In this case, NO_2 is involved in both steps of the mechanism and so the equation involves two terms; further simplification by substitution for the term $[NO_3]$ is required.

Clearly, in this case, equation 59 offers the easiest — that is, the simplest in mathematical terms — route to the chemical rate equation. In fact, this is a particularly straightforward example, but none the less it illustrates the general point. To analyse a mechanism it is best to concentrate on the theoretical rate equations that can be most easily simplified. We shall comment on this approach further when we analyse other mechanisms later in the Block.

The strategy that we have used in this Section to examine the kinetics of a composite reaction is not unique. As you will see time and time again in working through this Block, the first step of any kinetic analysis is to write down selected theoretical rate equations based on the elementary steps of the mechanism. If the reaction exhibits *time-independent* stoichiometry, we can then use the steady-state approximation to simplify such sets of, often unwieldy, expressions. If, however, the reaction exhibits *time-dependent* stoichiometry, then we have to use a different approach, and this is discussed in the next Section.

3.4 Summary of Section 3

1   One important method of testing a proposed mechanism is to compare the chemical rate equation predicted by the mechanism with the experimental rate equation.

2   The first step in any kinetic analysis of a mechanism is to determine the theoretical rate equations for the reactants, products and intermediates based on the elementary steps of the mechanism. A strategy for doing this is summarized in Box 3.

3   The theoretical rate equations obtained using the strategy in Box 3 are often unwieldy and difficult to simplify. One of the reasons for this is that we may be unable to determine the concentration of an intermediate at any given time during the reaction. This problem can sometimes be overcome by using the steady-state approximation, as outlined in Box 4.

but only if the intermediate is consumed almost as fast as it is produced, so its conc never becomes significant!

BOX 3   Strategy for deriving theoretical rate equations for a mechanism

- Give each elementary step a theoretical rate constant (k_1, k_{-1}, k_2, etc.) that describes its position in the sequence, and whether it is a forward or reverse reaction.

- From inspection of the steps, write down the theoretical rate equations. If a species is involved in more than one step, its overall rate of change in concentration can be obtained by summing the *individual rates of change in concentration* arising from each step when treated in isolation. *Remember to take account of stoichiometric numbers where necessary.* It can often be an advantage to concentrate on just those theoretical rate equations that can be most easily simplified.

BOX 4   The steady-state approximation

The steady-state approximation can be employed to simplify theoretical rate equations if a reaction exhibits time-independent stoichiometry — that is, if the concentration of the intermediate never becomes significant compared to that of the reactants and products. The assumption is that d[intermediate]/dt is equal to zero. However, this does not imply that the concentration of the intermediate does not change with time, but rather that the change is insignificant when compared to other terms in the rate equation for the intermediate.

By approximating d[intermediate]/dt to zero, the rate equation for the intermediate will also be equal to zero. This approximation often allows us to obtain an expression for the concentration of an intermediate in terms of the concentrations of the reactants and products. This expression can then be used to simplify other theoretical rate equations.

STUDY COMMENT   It is very important that you are able to write down theoretical rate equations for a proposed mechanism, and thence derive the chemical rate equation for a reaction that has time-independent stoichiometry. Try this out for yourself by working through SAQs 3 and 4.

SAQ 3   In Section 2.1.3 a two-step mechanism was proposed for the conversion of hypochlorite ion into chlorate ion:

$$2ClO^- \longrightarrow ClO_2^- + Cl^- \tag{37}$$

$$ClO_2^- + ClO^- \longrightarrow ClO_3^- + Cl^- \tag{38}$$

Write down the four theoretical rate equations based on the elementary steps of this mechanism to describe how the concentration of each of the species varies with time.

SAQ 4   The conversion of hypochlorite ion into chlorate ion exhibits time-independent stoichiometry. Using the theoretical rate equations obtained in SAQ 3, apply the steady-state approximation to the reactive intermediate, ClO_2^-, and obtain an expression for its concentration in terms of that of ClO^-. Confirm that we need to simplify only *one* of the theoretical rate equations obtained in SAQ 3 by showing that they can *all* be simplified to give a chemical rate equation that is consistent with the experimental rate equation:

$$J = k_R[ClO^-]^2$$

STUDY COMMENT   Now that you have completed Section 3, you should watch video band 3 (*Reaction Mechanisms*) before going on to Section 4.

[handwritten margin notes: don't forget $J = k[ClO^-]^2$ since two ClO^- have to collide! practise this skill!]

4 THE KINETICS OF A REACTION THAT EXHIBITS TIME-DEPENDENT STOICHIOMETRY

This Section is concerned with examining the kinetics of a reaction that exhibits time-dependent stoichiometry. The example we have chosen is the oxidation of nitrous acid by hydrogen peroxide (equation 6), which was introduced in Section 2:

$$HNO_2(aq) + H_2O_2(aq) = HNO_3(aq) + H_2O(l) \tag{6}$$

STUDY COMMENT   The main material of this Section is contained in Section 5 of the AV Booklet and the accompanying tape sequence: band 5 on audiocassette 2. You should work through this AV component now. In doing so, you should concentrate on the more general ideas that are presented rather than try to carry out any of the mathematical derivations for yourself. In particular, you should realize that a reaction that has time-dependent stoichiometry cannot be characterized by a single experimental rate equation, and so this raises the problem of how to test a proposed mechanism. The first few frames of the sequence provide a useful summary of some of the material we have already discussed in Sections 2 and 3 of this Block.

4.1 Further comments

So far, we have examined two methods of testing a mechanism using kinetic data. In Section 3 you saw that, provided the concentration of an intermediate remains very low during a reaction, the *steady-state approximation* can be used to simplify a set of theoretical rate equations. This allows a chemical rate equation to be obtained that can be compared directly with the experimental rate equation. The second method was introduced in the AV sequence you have just studied, where we discussed a reaction that cannot be characterized by a single experimental rate equation. In this case, integration of the theoretical rate equations gave a set of integrated rate equations which could be compared directly with the experimental data. It should be stressed that this latter method is *exact* — it involves *no* approximations — and it may be employed *irrespective* of whether or not the reaction exhibits *time-dependent* stoichiometry. But there is a problem. If a simple two-step mechanism generates a fairly complicated set of integrated rate equations, then you will appreciate that mechanisms involving many steps will give very complex rate equations. Integration is then usually difficult, and often it is impossible to find an exact integrated form.

Perhaps you can now see why the steady-state approximation is so important. Not only does it provide a *much* simpler analysis, but for a *large* number of composite reactions that exhibit time-independent stoichiometry it is the *only* effective method of testing a mechanism using kinetic data. Indeed, we spend most of the rest of this Block considering such examples.

But what do we do if the reaction exhibits time-dependent stoichiometry and the theoretical rate equations turn out to be too difficult to integrate? In such cases, integration can be carried out *numerically*, by means of a computer. In this process, the proposed rate equations (together with values of the rate constants) are fed into the computer, which then simulates the curves that describe the change in concentration of the various species with time. The values of the rate constants are usually obtained from independent studies, related systems, or simply guessed at. The rough estimates can then be adjusted from one simulation to the next to obtain the best fit to the experimental data.

This type of analysis is particularly important for composite mechanisms involving many steps, and is frequently applied to industrial processes, environmental problems (such as the production of photochemical smog, or the depletion of stratospheric ozone by halocarbons), and even in food chemistry. The combustion of petrol in a car engine requires more than thirty steps to explain the observations at temperatures below 1 300 K! Using a computer to perform the numerical integration of the 29 differential rate equations, it is possible to model the temperature — and pressure — history of the combustion almost perfectly: but many of the variables are interdependent!

4.2 Summary of Section 4 and the associated AV sequence

1 For composite reactions that exhibit time-dependent stoichiometry we cannot arbitrarily speak of the rate of reaction without stating explicitly the species whose concentration is changing.

2 Another method of testing a proposed mechanism using kinetics is to integrate the set of theoretical rate equations, and then examine whether the integrated rate equations are in accord with the experimental data.

3 The integration step described in point 2 above is frequently difficult. Hence:

- if the reaction exhibits time-independent stoichiometry, the steady-state approximation can be applied;

- if the reaction exhibits time-dependent stoichiometry, the integration process can usefully be carried out with a computer.

5 THE TRUE POWER OF KINETICS

5.1 An example: non-complementary electron transfer

In previous Sections we have used kinetic evidence in a confirmatory sense, checking that a probable mechanism predicts chemical rate equations, or that integrated rate equations are in accord with the experimental data. In this Section we shall demonstrate the true power of kinetics — how, given a number of possible mechanisms, it can weed out unlikely ones, so that we are left with a single probable pathway. It also provides us with an opportunity to apply the steady-state approximation to a more complex mechanism.

We shall take as our example the redox reaction between thallium(III), Tl^{3+}, and iron(II), Fe^{2+}:

$$Tl^{3+}(aq) + 2Fe^{2+}(aq) = Tl^+(aq) + 2Fe^{3+}(aq) \tag{79}$$

- Which ion is being oxidized and which reduced?

- The thallium gains two electrons, and is therefore being reduced, whereas the iron loses one electron and so is being oxidized.

Since the iron and the thallium atoms exchange different numbers of electrons in equation 79, we call this a **non-complementary electron transfer**. Strictly speaking, the four cations shown in the equation do not exist as free ions in solution. Rather, they form complexes, the nature of which depends on which anions are available to act as ligands. To discuss this reaction in terms of the various complexes would be to complicate an essentially *simple* transformation.

To overcome this problem we shall represent the various species present in terms of their oxidation states; that is

$$Tl^{III} + 2Fe^{II} = Tl^{I} + 2Fe^{III} \tag{80}$$

where, for example, Tl^{III} is taken to represent thallium in an oxidation state of +3.

The reaction exhibits *time-independent* stoichiometry, and the experimentally determined rate equation has the form

$$J = k_R[Tl^{III}][Fe^{II}] \tag{81}$$

■ Is this reaction likely to proceed via a composite mechanism?

▨ Yes. If this reaction proceeded via a single elementary step, it would be termolecular, which is very unlikely. Similarly, from inspection of the stoichiometry it would be expected to have the following chemical rate equation:

$$J = k_R[Tl^{III}][Fe^{II}]^2 \qquad \leftarrow \qquad note! \tag{82}$$

Since the experimental rate equation does not take this form, it *must* proceed via a composite mechanism.

Having established that the reaction is composite, the next stage is to propose possible mechanisms that are consistent with the criteria outlined in Section 2.1.2. The simplest and most likely mechanism will involve two consecutive steps.

STUDY COMMENT   Before reading further, spend a few moments seeing how many two-step composite mechanisms you can write down for this reaction, each step involving the exchange of one or two electrons. At this stage, don't worry about trying to accommodate any of the kinetic data or about having to include unusual oxidation numbers of iron or thallium. Just ensure that the mechanisms you propose add up to the stoichiometric equation, and involve only unimolecular and bimolecular steps.

The three most likely mechanisms are:

Composite Mechanism A

$$Tl^{III} + Fe^{II} \longrightarrow \mathbf{Tl^{II}} + Fe^{III} \tag{83}$$
$$\mathbf{Tl^{II}} + Fe^{II} \longrightarrow Tl^{I} + Fe^{III} \tag{84}$$

Composite Mechanism B

$$Tl^{III} + Fe^{II} \longrightarrow Tl^{I} + \mathbf{Fe^{IV}} \tag{85}$$
$$\mathbf{Fe^{IV}} + Fe^{II} \longrightarrow 2Fe^{III} \tag{86}$$

Composite Mechanism C

$$Fe^{II} + Fe^{II} \longrightarrow Fe^{III} + \mathbf{Fe^{I}} \tag{87}$$
$$\mathbf{Fe^{I}} + Tl^{III} \longrightarrow Tl^{I} + Fe^{III} \tag{88}$$

In each case, two of the reactant species combine in the first step to give one of the products plus an intermediate ($\mathbf{Tl^{II}}$, $\mathbf{Fe^{IV}}$ and $\mathbf{Fe^{I}}$, respectively, shown here and in the remainder of Section 5 in bold type). The second step involves the reaction of the intermediate with another reactant species to give one, or both, product species.

Other mechanisms are possible, but these all involve more than two steps. So, placing our faith in the worthy William of Occam, we shall concentrate on the simple two-step mechanisms above.

■ What criteria can we use to weed out the unsuitable candidates from these three mechanisms?

■ Looking back to Section 2.1.2, the mechanism should accommodate all the available data. We know the form of the experimental rate equation (equation 81), so an obvious test is to check which of the mechanisms predict(s) chemical rate equations that are in agreement with this expression. The mechanism should also be chemically reasonable; we may be able to discard mechanisms on the basis of unlikely oxidation numbers of thallium or iron.

There is insufficient information on the stabilities of the oxidation states of thallium or iron to say *categorically* that one mechanism is much more likely than another. Analysis of the kinetics provides a more satisfactory method of distinguishing the most probable pathway, and this is the approach that we shall employ.

■ In Section 2 we deduced the mechanism for the oxidation of nitrous acid to nitric acid by detecting and identifying the intermediate. Why would it be difficult to do this in this case?

▨ Since the reaction exhibits time-independent stoichiometry, the concentration of the reactive intermediate will be very low, and thus difficult to detect.

5.2 A kinetic analysis of the three possible mechanisms

Before we attempt to derive the theoretical rate equations, some modification of the mechanisms is necessary. If the reaction exhibits time-independent stoichiometry, then, as stated earlier, the intermediate must be present in very low concentrations. This situation arises for one of two reasons:

1   The second step in a given mechanism is much faster than the first, so that the intermediate reacts to give the products almost as soon as it is formed. (The reaction between N_2O_5 and NO that you met in Section 2 fits into this category.)

2   The first step does not go 'to completion', but is 'reversible': if the equilibrium position favours the reactants, then the concentration of the intermediate will *always* be relatively small.

STUDY COMMENT   Try the following SAQ so that you can see how the equilibrium constant is related to the rate constants for the forward and reverse reactions for a simple reaction whose mechanism consists of two opposing elementary steps.

SAQ 5
(a)   Write down the theoretical rate equations for the reactant, A, and the product, B, for a reversible reaction whose mechanism consists of just two opposing elementary steps:

$$A \underset{k_{-1}}{\overset{k_1}{\rightleftarrows}} B \qquad\qquad (89)$$

(b)   What is the rate of change in concentration of A and B when this reaction is at equilibrium?

(c)   Assuming this reaction is at equilibrium, derive an expression for the ratio k_1/k_{-1} in terms of the equilibrium concentrations, $[A]_{eq}$ and $[B]_{eq}$. What does this tell you about the relationship between the rate constants k_1 and k_{-1} and the equilibrium constant for the reaction, K_c?

We know from SAQ 5 that K_c, the equilibrium constant for opposing elementary steps, is simply the ratio of the individual rate constants for the forward and reverse reactions, that is

$$K_c = \frac{k(\text{forward})}{k(\text{reverse})} \qquad\qquad (90)$$

Inspection of the magnitude of the equilibrium constant should therefore reveal whether the reverse reaction occurs to any significant extent.

An estimate of the equilibrium constant for the first step in Mechanism A (equation 83) gives a value of about 4×10^{-8}, which suggests that the reverse reaction *is likely to be important*, and in the absence of evidence to the contrary, should be included in the mechanism. As we shall see later, this reverse reaction is in fact *very important*. Similar arguments can be put forward for the other mechanisms, so they should all be rewritten as:

Composite Mechanism A

$$Tl^{III} + Fe^{II} \underset{k_{-1}}{\overset{k_1}{\rightleftarrows}} \mathbf{Tl^{II}} + Fe^{III} \tag{91}$$

$$\mathbf{Tl^{II}} + Fe^{II} \xrightarrow{k_2} Tl^{I} + Fe^{III} \tag{92}$$

Composite Mechanism B

$$Tl^{III} + Fe^{II} \underset{k_{-1}}{\overset{k_1}{\rightleftarrows}} Tl^{I} + \mathbf{Fe^{IV}} \tag{93}$$

$$\mathbf{Fe^{IV}} + Fe^{II} \xrightarrow{k_2} 2\,Fe^{III} \tag{94}$$

Composite Mechanism C

$$Fe^{II} + Fe^{II} \underset{k_{-1}}{\overset{k_1}{\rightleftarrows}} Fe^{III} + \mathbf{Fe^{I}} \tag{95}$$

$$\mathbf{Fe^{I}} + Tl^{III} \xrightarrow{k_2} Tl^{I} + Fe^{III} \tag{96}$$

We are now in a position to derive the chemical rate equations predicted by these three mechanisms. Let's start with composite Mechanism A.

As before, the first step is to write down the theoretical rate equations for reactants, intermediates and products based on the elementary steps of the mechanism. For the moment, we shall not try to simplify matters by being selective in this process.

■ Write down the rate equation for the product Tl^{I} according to Mechanism A.

▢ Tl^{I} is formed in step 92 alone. Hence,

$$\frac{d[Tl^{I}]}{dt} = k_2[\mathbf{Tl^{II}}][Fe^{II}] \tag{97}$$

■ Write down the theoretical rate equation for the reactive intermediate $\mathbf{Tl^{II}}$ according to Mechanism A.

▢ $\mathbf{Tl^{II}}$ is formed in the forward reaction of step 1 (equation 91), lost in the reverse reaction of step 1 (equation 91), and also lost in the second step (equation 92). Thus, in detail,

$$\left(\frac{d[\mathbf{Tl^{II}}]}{dt}\right)_1 = k_1[Tl^{III}][Fe^{II}] \tag{98}$$

$$\left(\frac{d[\mathbf{Tl^{II}}]}{dt}\right)_{-1} = -k_{-1}[\mathbf{Tl^{II}}][Fe^{III}] \tag{99}$$

$$\left(\frac{d[\mathbf{Tl^{II}}]}{dt}\right)_2 = -k_2[\mathbf{Tl^{II}}][Fe^{II}] \tag{100}$$

Combining equations 98, 99 and 100, gives

$$\frac{d[\mathbf{Tl^{II}}]}{dt} = k_1[Tl^{III}][Fe^{II}] - k_{-1}[\mathbf{Tl^{II}}][Fe^{III}] - k_2[\mathbf{Tl^{II}}][Fe^{II}] \tag{101}$$

(Note that this *total* expression could have been written down in a single step by inspection of the mechanism.)

Equation 101 can be tidied up to give the following expression:

$$\frac{d[\mathbf{Tl^{II}}]}{dt} = k_1[Tl^{III}][Fe^{II}] - \{k_{-1}[Fe^{III}] + k_2[Fe^{II}]\}[\mathbf{Tl^{II}}] \tag{102}$$

We can obtain similar theoretical rate equations for Tl^{III}, Fe^{II} and Fe^{III}:

$$\frac{d[Tl^{III}]}{dt} = -k_1[Tl^{III}][Fe^{II}] + k_{-1}[\mathbf{Tl^{II}}][Fe^{III}] \tag{103}$$

$$\frac{d[Fe^{II}]}{dt} = -k_1[Tl^{III}][Fe^{II}] + k_{-1}[\mathbf{Tl^{II}}][Fe^{III}] - k_2[\mathbf{Tl^{II}}][Fe^{II}] \tag{104}$$

$$\frac{d[Fe^{III}]}{dt} = k_1[Tl^{III}][Fe^{II}] - k_{-1}[\mathbf{Tl^{II}}][Fe^{III}] + k_2[\mathbf{Tl^{II}}][Fe^{II}] \tag{105}$$

These expressions are unwieldy and, in their present form, cannot be manipulated further to give a simple solution.

■ How can this set of theoretical rate equations be simplified?

▨ Since this reaction exhibits time-independent stoichiometry, we can apply the steady-state approximation to the intermediate $\mathbf{Tl^{II}}$; that is, assume

$$\frac{d[\mathbf{Tl^{II}}]}{dt} = 0 \tag{106}$$

This assumption means that equation 102 becomes

$$0 = k_1[Tl^{III}][Fe^{II}] - \{k_{-1}[Fe^{III}] + k_2[Fe^{II}]\}[\mathbf{Tl^{II}}] \tag{107}$$

From this we can obtain an expression for the concentration of the intermediate, $\mathbf{Tl^{II}}$:

$$[\mathbf{Tl^{II}}] = \frac{k_1[Tl^{III}][Fe^{II}]}{\{k_{-1}[Fe^{III}] + k_2[Fe^{II}]\}} \tag{108}$$

This expression can be used to simplify any of the theoretical rate equations 97, 103, 104 and 105.

■ Which one of these equations would you choose to simplify?

▨ The most straightforward is equation 97, since it is the only one that involves a single term.

Substituting the expression for $[\mathbf{Tl^{II}}]$ into equation 97 gives a rate equation for the product, Tl^{I}, in terms of just the concentrations of reactants and products; that is

$$\frac{d[Tl^{I}]}{dt} = \frac{k_1 k_2[Tl^{III}][Fe^{II}]^2}{\{k_{-1}[Fe^{III}] + k_2[Fe^{II}]\}} \tag{109}$$

The stoichiometric equation

$$Tl^{III} + 2Fe^{II} = Tl^{I} + 2Fe^{III} \tag{80}$$

can be used to express the overall rate of reaction, J, in terms of the product, Tl^{I}:

CHEMICAL RATE EQUATION PREDICTED BY MECHANISM A

$$J = \frac{d[Tl^{I}]}{dt} = \frac{k_1 k_2[Tl^{III}][Fe^{II}]^2}{\{k_{-1}[Fe^{III}] + k_2[Fe^{II}]\}} \tag{110}$$

STUDY COMMENT Take one of the remaining theoretical rate equations (103, 104 or 105), and substitute for [Tl^{II}] the expression given by equation 108. You should find that it is then possible to rearrange your expression to give the chemical rate equation as expressed by equation 110. However, a series of manipulations will be required! This highlights the advantage of selecting the theoretical rate equation — for either a reactant or product — that will be the 'easiest to deal with' at the outset of a kinetic analysis of a mechanism. Do find the time to try this.

As it stands, equation 110 doesn't look much like the experimental expression in equation 81:

$$J = k_R[Tl^{III}][Fe^{II}] \tag{81}$$

Notice, however, that equation 110 can be rewritten as:

$$J = \left\{ \frac{k_1 k_2 [Fe^{II}]}{(k_{-1}[Fe^{III}] + k_2[Fe^{II}])} \right\} \times [Tl^{III}][Fe^{II}] \tag{111}$$

■ How can the first term (that in curly brackets) be made into a constant, so that equation 111 can be reduced to a form similar to equation 81? (Consider the relative magnitudes of the terms in the denominator.)

▨ If the term $k_2[Fe^{II}]$ is much greater than the term $k_{-1}[Fe^{III}]$ (that is, $k_2[Fe^{II}] \gg k_{-1}[Fe^{III}]$), it will dominate the denominator, and the equation then reduces to:

$$J = \left\{ \frac{k_1 k_2 [Fe^{II}]}{k_2[Fe^{II}]} \right\} \times [Tl^{III}][Fe^{II}] = k_1[Tl^{III}][Fe^{II}] \tag{112}$$

This now has the same form as the experimental rate equation 81 ($k_1 = k_R$).

(The physical significance of assuming that the term $k_2[Fe^{II}]$ dominates the denominator in equation 111 is discussed in more general terms in Section 6 of the AV Booklet.)

Our conclusion is that Mechanism A *can* generate a chemical rate equation that is consistent with the rate equation that is observed experimentally. So, on this evidence we cannot discount Mechanism A.

STUDY COMMENT You should now try the following SAQ. It gives you the opportunity to derive for yourself the chemical rate equation predicted by Mechanism C.

SAQ 6

Composite Mechanism C

$$Fe^{II} + Fe^{II} \underset{k_{-1}}{\overset{k_1}{\rightleftharpoons}} Fe^{III} + Fe^{I} \tag{95}$$

$$Fe^{I} + Tl^{III} \xrightarrow{k_2} Tl^{I} + Fe^{III} \tag{96}$$

(a) Write down the theoretical rate equations for (i) Tl^{I} and (ii) **Fe^{I}** based on the elementary steps of this mechanism.

(b) Apply the steady-state approximation to the intermediate, **Fe^{I}**, and thus obtain an expression for its concentration in terms of those of the reactants and products.

(c) Use this equation to obtain the theoretical rate equation for Tl^{I}, and thus the chemical rate equation, in terms of just the reactants and products.

In summary, according to the analysis in SAQ 6:

CHEMICAL RATE EQUATION PREDICTED BY MECHANISM C

$$J = \frac{d[Tl^I]}{dt} = \frac{k_1 k_2 [Tl^{III}][Fe^{II}]^2}{k_{-1}[Fe^{III}] + k_2[Tl^{III}]} = \left\{ \frac{k_1 k_2 [Fe^{II}]}{(k_{-1}[Fe^{III}] + k_2[Tl^{III}])} \right\} \times [Tl^{III}][Fe^{II}] \quad (113)$$

▪ Is there any way in which equation 113 can be simplified into a form similar to equation 81; that is, $J = k_R[Tl^{III}][Fe^{II}]$?

▪ No. Even if one term in the denominator predominates, the expression in curly brackets does *not* reduce to a constant, and so the equation *cannot* be manipulated into a form similar to equation 81.

This rules out Mechanism C as a likely candidate.

Finally, the chemical rate equation predicted by Mechanism B can be obtained in just the same way as those for Mechanisms A and C. It takes the following form:

CHEMICAL RATE EQUATION PREDICTED BY MECHANISM B

$$J = \frac{k_1 k_2 [Tl^{III}][Fe^{II}]^2}{\{k_{-1}[Tl^I] + k_2[Fe^{II}]\}} = \left\{ \frac{k_1 k_2 [Fe^{II}]}{(k_{-1}[Tl^I] + k_2[Fe^{II}])} \right\} \times [Tl^{III}][Fe^{II}] \quad (114)$$

In this case, if the term $k_2[Fe^{II}]$ is much greater than $k_{-1}[Tl^I]$ (that is, $k_2[Fe^{II}] \gg k_{-1}[Tl^I]$), it will dominate the denominator, and the equation reduces to

$$J = k_1[Tl^{III}][Fe^{II}] \quad (115)$$

This now has the same form as the experimental rate equation 81 ($k_1 = k_R$).

▪ From a consideration of the chemical rate equations predicted by Mechanisms A, B and C and their reduced forms, which of the mechanisms is most likely?

▪ The only mechanism that generates a chemical rate equation that is *not* consistent with the experimental rate equation 81 is Mechanism C. Therefore, at this stage, Mechanisms A and B are equally likely.

▪ If you look at the chemical rate equations 110 and 114 predicted by Mechanisms A and B, you will see that they both contain the concentration terms of the two reactants, $[Tl^{III}]$ and $[Fe^{II}]$, but only one product concentration appears in each equation — $[Fe^{III}]$ in equation 110 and $[Tl^I]$ in equation 114. How could we make use of this information to differentiate between the two proposed mechanisms?

▪ If a high concentration of one of these products is added to the reaction mixture, we would expect the reaction rate to be affected if the rate equation contains a concentration term for this species.

▪ In fact, when a high concentration of the product Fe^{III} is added to the reaction, the rate of reaction decreases. From a consideration of the two chemical rate equations 110 and 114, try to decide which of the two mechanisms, A or B, is more likely.

▪ The concentration of Fe^{III} does not appear in equation 114. Thus, if the reaction proceeds via Mechanism B, addition of a high concentration of Fe^{III} should not affect the rate. By contrast, the term $[Fe^{III}]$ *does* appear in the chemical rate equation for Mechanism A, equation 110. Adding a large concentration of Fe^{III} could make the term $k_{-1}[Fe^{III}]$ of similar magnitude to $k_2[Fe^{II}]$, such that the latter term *no longer dominates* the denominator. The rate of reaction would thus depend on $[Fe^{III}]$; the rate would be *decreased* when $[Fe^{III}]$ was increased, since this term appears in the denominator of equation 110. Thus, Mechanism A is the more likely.

By studying the dependence of the rate of reaction on the concentration of one of the products, we have been able to differentiate between Mechanisms A and B, leaving Mechanism A as the more probable pathway. However, you should always remember that although this is the most likely mechanism on present evidence, there is still the possibility that the reaction proceeds via another mechanism that we have not yet considered.

The material in this Section has demonstrated an important aspect of chemical kinetics, namely how a comparison of the chemical rate equation predicted by a number of possible mechanisms allows us to eliminate unlikely ones. In the next Section (and the AV sequence associated with it) we shall take up a point that was touched on during our manipulation of equation 111. To reduce a chemical rate equation to a form comparable with that found experimentally (for example, in going from equation 111 to equation 112), it is often necessary to assume that one term dominates the denominator. But what is the *physical significance* of this type of assumption, and what does it mean for the mechanism under study?

5.3 Summary of Section 5

1 There are three possible two-step mechanisms that could account for the non-complementary electron transfer between Tl^{III} and Fe^{II}.

2 By applying the steady-state approximation (Box 4) to the intermediates involved in these mechanisms, it is possible to derive the chemical rate equation for each of the three mechanisms. This analysis is simplified by selecting a theoretical rate equation — for either a reactant or a product — which keeps any mathematical manipulation to a minimum. Of the three chemical rate equations, two can be reduced to a form compatible with the experimental rate equation.

3 By studying how the addition of a high concentration of one of the products, Fe^{III}, affects the value of the observed rate, it is possible to distinguish between the remaining two mechanisms on the basis of their chemical rate equations.

STUDY COMMENT   Now try the following SAQ. It will help you to consolidate your skills of deriving chemical rate equations using the steady-state approximation, and then, using the technique used in Section 5.2, you can distinguish between the proposed mechanisms. This would also be a good point to watch video band 3 if you have not already done so.

SAQ 7   The reaction between nitric oxide, NO, and oxygen, O_2, in the gas phase,

$$2NO(g) + O_2(g) = 2NO_2(g) \tag{116}$$

exhibits time-independent stoichiometry and has the following experimental rate equation:

$$J = k_R[NO]^2[O_2] \tag{117}$$

Two possible pathways for this reaction are:

a two-step mechanism

$$NO + O_2 \underset{k_{-1}}{\overset{k_1}{\rightleftharpoons}} NO_3 \tag{118}$$

$$NO_3 + NO \xrightarrow{k_2} 2NO_2 \tag{119}$$

or *a single elementary step**

$$2NO + O_2 \xrightarrow{k_1'} 2NO_2 \tag{120}$$

* The use of the prime in the rate constant k_1' for equation 120 simply indicates that its numerical value is different from that of k_1 in equation 118.

(a)   For the two-step mechanism, write down the theoretical rate equations for (i) NO_3 and (ii) NO_2 based on the elementary steps of this mechanism. By applying the steady-state approximation to the intermediate, NO_3, in this mechanism, derive the theoretical rate equation for NO_2, and thus the chemical rate equation, in terms of the concentrations of NO and O_2 alone. Does this agree with the experimental rate equation?

(b)   What are the advantages of substituting for $[NO_3]$ into the theoretical rate equation for NO_2 over substituting into those for NO or O_2?

(c)   What is the chemical rate equation for the elementary step 120? Does kinetics help us to distinguish between these two mechanisms?

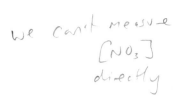

we can't measure $[NO_3]$ directly

6 RATE-LIMITING STEPS AND PRE-EQUILIBRIA

In the last Section you saw that the chemical rate equation derived from a proposed mechanism was compatible with the experimental rate equation only if it was assumed that the relative magnitudes of certain terms in the derived expression were such that one term was dominant. It is important to look at this strategy in more detail and, in particular, to examine the consequences that this type of assumption has for a given mechanism.

STUDY COMMENT   The material contained in Section 6 of the AV Booklet (*Rate-limiting steps and pre-equilibria*) and the accompanying tape sequence (band 6 on audiocassette 2) forms an integral part of the development in this Section. You should work through this AV sequence now. Again, you should concentrate on the general ideas that are presented rather than try to carry out any detailed mathematical derivations for yourself. In particular, you should notice how assumptions that can be made about the relative rates of the steps in a mechanism can be given a distinct physical meaning. When you have finished the sequence, try to work through SAQ 8.

SAQ 8   The first mechanism you met in the AV sequence is repeated below:

$$A \xrightarrow{\ k_1\ } B \qquad \qquad (121)$$

slow (low k_1)

$$B \xrightarrow{\ k_2\ } C \qquad \qquad (122)$$

fast (high k_2)

When $k_2 \gg k_1$, the reaction exhibits time-independent stoichiometry. Apply the steady-state approximation to the intermediate, B, and thus simplify the theoretical rate equations to give a chemical rate equation containing [A] alone. Is this expression in accord with the conclusions we came to in the AV sequence?

As you should have found in working through the AV sequence, a chemical rate equation predicted by a mechanism can often be reduced to a simpler expression, consistent with its experimental counterpart, by assuming that one term in the equation dominates. This can usually be interpreted in terms of one step being intrinsically much slower than another, an assumption that often allows us to simplify the mechanism, as well as the rate equations.

As a concrete example, consider again Mechanism A for the non-complementary electron transfer discussed in Section 5:

$$Tl^{III} + Fe^{II} \underset{k_{-1}}{\overset{k_1}{\rightleftharpoons}} Tl^{II} + Fe^{III} \qquad \qquad (91)$$

$$Tl^{II} + Fe^{II} \xrightarrow{\ k_2\ } Tl^{I} + Fe^{III} \qquad \qquad (92)$$

As you saw there, the mechanism predicts the following chemical rate equation:

$$J = \frac{d[Tl^I]}{dt} = \frac{k_1 k_2 [Tl^{III}][Fe^{II}]^2}{\{k_{-1}[Fe^{III}] + k_2[Fe^{II}]\}} \tag{110}$$

If it is assumed that the term $k_{-1}[Fe^{III}]$ is much smaller than $k_2[Fe^{II}]$, then this expression is reduced to a form consistent with the experimental rate equation. What does this tell us about the mechanism of the reaction?

If $k_2[Fe^{II}] \gg k_{-1}[Fe^{III}]$, equation 110 reduces to

$$J = k_1[Tl^{III}][Fe^{II}] \tag{112}$$

The right-hand side of this equation is identical to the right-hand side of the rate equation for the first step in the forward direction (equation 98 in Section 5.2). This is analogous to one of the limiting cases we discussed in the AV sequence. The term $k_2[Fe^{II}]$ relates to the rate of the second step, whereas $k_{-1}[Fe^{III}]$ relates to the reverse reaction of the first step. Since $k_2[Fe^{II}]$ is much greater than $k_{-1}[Fe^{III}]$, this means that virtually all of the intermediate Tl^{II} goes on to give the product. The reverse reaction thus becomes unimportant, and the mechanism effectively reduces to *two irreversible consecutive steps*. Since Tl^{II} is a reactive intermediate, it reacts almost as soon as it is formed, and the first step is the **rate-limiting step**, as indicated by equation 112.

Of course, in general terms we still need to include the reverse reaction. For example, without it we would not have been able to explain the decrease in the rate on adding a large concentration of the product, Fe^{III}. In such circumstances, $k_2[Fe^{II}]$ is no longer *much* greater than $k_{-1}[Fe^{III}]$; the *reverse reaction is then no longer unimportant*.

What would equation 110 reduce to if $k_{-1}[Fe^{III}]$ were much greater than $k_2[Fe^{II}]$? What would this tell us about the mechanism?

If $k_{-1}[Fe^{III}] \gg k_2[Fe^{II}]$, equation 110 reduces to

$$J = \frac{k_1 k_2}{k_{-1}} \times \frac{[Tl^{III}][Fe^{II}]^2}{[Fe^{III}]} \tag{123}$$

We can understand the form of this equation in more detail if we explore what happens when we assume that the first step in the mechanism (equation 91) can be treated in isolation. In this case the equilibrium constant for this step is

$$K_c = \frac{k_1}{k_{-1}} = \frac{[Tl^{II}][Fe^{III}]}{[Tl^{III}][Fe^{II}]} \tag{124}$$

and so the equilibrium concentration of Tl^{II} can be expressed as

$$[Tl^{II}] = \frac{k_1}{k_{-1}} \times \frac{[Tl^{III}][Fe^{II}]}{[Fe^{III}]} \tag{125}$$

By substituting this expression for $[Tl^{II}]$ into equation 97

$$\frac{d[Tl^I]}{dt} = k_2[Tl^{II}][Fe^{II}] \tag{97}$$

we have

$$J = \frac{d[Tl^I]}{dt} = \left\{ \frac{k_1}{k_{-1}} \times \frac{[Tl^{III}][Fe^{II}]}{[Fe^{III}]} \right\} \times k_2[Fe^{II}] \tag{126}$$

This expression is identical to that in equation 123. So, for the condition $k_{-1}[Fe^{III}] \gg k_2[Fe^{II}]$, we can treat the first step in the mechanism as a **pre-equilibrium**. In physical terms, this means that a given ion of Tl^{II} has 'more chance' of being re-oxidized to Tl^{III} than of reacting with Fe^{II} to give products. It also means that it is the *second step* in the mechanism that we regard as being rate limiting.

The idea of a rate-limiting step provides an invaluable technique for simplifying the potentially complicated mathematical analysis of some composite mechanisms. It can also be very useful in the analysis of more straightforward mechanisms if there are grounds for assuming that a particular step is relatively slow. A common strategy for reactions that exhibit *time-independent stoichiometry* is given in Box 5.

BOX 5   A strategy for simplifying the analysis of composite reaction mechanisms that exhibit time-independent stoichiometry, using the concept of a rate-limiting step

1   Propose a mechanism that is consistent with the criteria outlined in Section 2.1.2.

2   If analysis using the steady-state approximation is complex, assume one step is rate limiting, such that the *overall* rate of reaction is equivalent to the rate of this step. *In other words, J for the overall reaction can be set equal to the rate equation of the rate-limiting step.* * This procedure can also be followed if there are grounds for believing that a particular step in a mechanism is relatively slow.

3   Simplify the resulting rate equation by assuming that any steps *prior* to the rate-limiting step are rapidly established pre-equilibria. This facilitates the calculation of the concentration of any intermediate species formed before the rate-limiting step, since their concentrations can be taken to be the equilibrium values, as determined by the equilibrium step in which they are formed. (In order that the intermediates do not build up to substantial concentrations, the equilibrium constants of these steps must favour the reactants; that is, $K_c \ll 1$.) *Steps after the rate-limiting step do not have any effect on the overall rate of reaction.*

4   Compare the chemical rate equation generated in stage 3 with the experimental rate equation. If they differ, it may be possible to obtain another chemical rate equation by assuming that a different step is rate limiting.

As you can see, deciding which step is rate limiting is frequently a matter of trial and error, although it is sometimes made a bit easier if the decision can be based on some chemical criterion. *However, in this Course you will not be expected to decide which step is rate limiting, but will always be given this information.*

STUDY COMMENT   Given a reaction with time-independent stoichiometry and a proposed reaction mechanism, you should by now be familiar with determining the chemical rate equation using the steady-state approximation. However, it is equally important that you have experience in deriving a chemical rate equation when it is known that a particular step in a reaction mechanism — for a reaction that has time-independent stoichiometry — is rate limiting. The key point in this case is that it is not necessary to develop an analysis in which the steady-state approximation is used; if you like, a 'short cut' is available. To see how this short cut works, make sure you try the following SAQ before moving on.

SAQ 9   In Block 2 you analysed the iodide–hypochlorite reaction (equation 1), which exhibits time-independent stoichiometry:

$$I^-(aq) + ClO^-(aq) = IO^-(aq) + Cl^-(aq) \tag{1}$$

This has been found to have the following experimental rate equation:

$$J = k_R \frac{[I^-][ClO^-]}{[OH^-]} \tag{127}$$

* This is valid only if the rate-limiting step occurs once for each occurrence of the reaction mechanism. In other words, the sum of the various steps in the mechanism must add up to the stoichiometric equation *without* including the rate-limiting step more than once.

The mechanism proposed for this reaction is as follows

$$ClO^- + H_2O \underset{k_{-1}}{\overset{k_1}{\rightleftharpoons}} HClO + OH^- \qquad rapid \quad (128)$$

$$HClO + I^- \xrightarrow{k_2} HIO + Cl^- \qquad slow \quad (129)$$

$$OH^- + HIO \xrightarrow{k_3} H_2O + IO^- \qquad rapid \quad (130)$$

where the second step is rate limiting. Using the strategy outlined in Box 5, write down an expression for *J* for this reaction, based on the second step. Assuming that the first step is a rapidly established pre-equilibrium, write down an expression for the concentration of the intermediate HClO, and thus convert the chemical rate equation derived earlier into a form compatible with equation 127.

Having worked through SAQ 9, you should be able to appreciate that if the rates of steps *after* the rate-limiting step can be ignored (since they are fast compared to the rate of the limiting step), then any rate constant or species appearing in the mechanism *after* the rate-limiting step *does not* appear in the chemical rate equation.

6.1 Summary of Section 6 and the associated AV sequence

1 For a mechanism consisting of two first-order irreversible steps, if one rate constant is much smaller than the other, the rate of formation of the product is governed by the rate of the slow step, known as the rate-limiting step. This conclusion can be reached using both a qualitative and a quantitative analysis, and can be generalized to mechanisms consisting of any number of such steps.

For the following two-step consecutive mechanism:

$$A \xrightarrow{k_1} B \xrightarrow{k_2} C$$

● if $k_2 \gg k_1$, the first step is rate limiting and the intermediate B is reactive. This would result in time-*independent* stoichiometry;

● if $k_1 \gg k_2$, the second step is rate limiting and the intermediate B is 'stable' (that is, less reactive). This would result in time-*dependent* stoichiometry.

2 For the following mechanism, where B is a reactive intermediate:

$$A \underset{k_{-1}}{\overset{k_1}{\rightleftharpoons}} B \xrightarrow{k_2} C$$

● if $k_2 \gg k_{-1}$, the reverse reaction becomes unimportant, and the mechanism reduces to two consecutive irreversible steps, the first of which is rate limiting; that is,

$$A \xrightarrow{slow} B \xrightarrow{fast} C$$

● if $k_{-1} \gg k_2$, the first step can be treated as a rapidly established equilibrium, a pre-equilibrium, the concentration of the intermediate being determined by the equilibrium constant for this step. In this case, the second step can be classed as rate limiting; that is,

$$A \underset{pre\text{-}equilibrium}{\rightleftharpoons} B \underset{\substack{slow, \\ rate\ limiting}}{\xrightarrow{k_2}} C$$

3 For reactions that exhibit time-independent stoichiometry, quite complex reaction pathways can be greatly simplified by assuming that one step in a mechanism is rate limiting (that is, *J* can be set equal to the rate equation of this step), and that all steps preceding the rate-limiting step are rapidly established equilibria.

STUDY COMMENT   The following SAQ gives you a further opportunity to practise generating chemical rate equations using the rate-limiting strategy outlined in Box 5.

SAQ 10   Assume now that the last step of the mechanism given in SAQ 9 is rate limiting; that is,

$$ClO^- + H_2O \underset{k_{-1}}{\overset{k_1}{\rightleftharpoons}} HClO + OH^- \qquad\qquad rapid \quad (128)$$

$$HClO + I^- \underset{k_{-2}}{\overset{k_2}{\rightleftharpoons}} HIO + Cl^- \qquad\qquad rapid \quad (131)$$

$$OH^- + HIO \xrightarrow{\;k_3\;} H_2O + IO^- \qquad\qquad slow \quad (132)$$

Use the strategy outlined in Box 5 to derive the chemical rate equation predicted by this mechanism in terms of $[ClO^-]$, $[I^-]$, $[H_2O]$ and $[Cl^-]$. Is this consistent with the experimental rate equation (equation 127 in SAQ 9)?

7 CHAIN REACTIONS

7.1 Introduction

One of the most important classes of composite reaction is the **chain reaction**. Such reactions occur in internal combustion engines, rockets, gas fires and gas cookers. They form the basis of many industrial processes such as polymerization, halogenation and atmospheric reactions, including the formation of photochemical smog and the maintenance of the stratospheric ozone layer. In this Section we shall examine one such reaction, the gas-phase thermal decomposition of ethanal, CH_3CHO (equation 133), which outlines the essential ingredients of a chain reaction:

$$CH_3CHO(g) = CH_4(g) + CO(g) \qquad\qquad (133)$$

At 753 K this reaction exhibits time-independent stoichiometry, and has the following experimental rate equation:

$$J = k_R[CH_3CHO]^{3/2} \qquad\qquad (134)$$

■   Does this reaction proceed via a composite mechanism under these conditions?

▨   Yes; if it occurred via a single elementary step, the rate equation would have the form

$$J = k_R[CH_3CHO] \qquad\qquad (135)$$

7.2 The mechanism of a chain reaction

Employing the strategy developed in Section 2.1, the next task is to propose a mechanism that is consistent with the criteria set out in Section 2.1.2. However, without a knowledge of the chemical background of this type of reaction, it would be impossible for you to draw up a suitable pathway. We shall therefore give you the most likely solution. For many organic compounds in the gas phase, heating results in cleavage at the weakest bond in the molecule to give two 'radicals'. A possible mechanism is

$$CH_3CHO \xrightarrow{\;k_1\;} CH_3\cdot + CHO\cdot \qquad\qquad\qquad initiation \quad (136)$$

$$CH_3\cdot + CH_3CHO \xrightarrow{\;k_2\;} CH_4 + CH_3CO\cdot \qquad\qquad propagation \quad (137)$$

$$CH_3CO\cdot \xrightarrow{\;k_3\;} CH_3\cdot + CO \qquad\qquad\qquad propagation \quad (138)$$

$$2CH_3\cdot \xrightarrow{\;k_4\;} C_2H_6 \qquad\qquad\qquad termination \quad (139)$$

The main justification for employing this mechanism is that it seems to fit the facts.*

The species $CH_3\cdot$, $CHO\cdot$ and $CH_3CO\cdot$ are **radicals**, and are very reactive: they can be described as 'molecular fragments which have an unpaired electron'. Since radicals have an odd number of electrons, any reaction with a 'normal' molecule (that is, usually one with all of the electrons paired), as shown in equations 137 and 138, *must* produce another radical. Radicals are formed initially in the first step of the mechanism, which is known as the **initiation reaction**.

■   What is the sum of the second and third steps in this mechanism — that is, those labelled 'propagation'?

▨   The sum of these two steps is the stoichiometric equation (133).

Notice that the methyl radical, $CH_3\cdot$, which is a reactant in the second step, is regenerated in the third step. This means it can now react with another molecule of ethanal as in the second step, and duplicate the process to form more methane, CH_4, and carbon monoxide, CO. This cycle repeats itself many times in the sequence (137, 138, 137, 138, 137,...). So far in this Block we have only considered *open sequence* reaction mechanisms, in which the intermediates formed in one step disappear in the next. A chain reaction as described above is an example of a *closed sequence*, in which an intermediate can be recycled: it is consumed in one step but re-formed in another. Processes such as the second and third steps of the above mechanism, which give rise to the product and continue the chain, are called the **propagation reactions**. Although they are depicted only once, you should realize that they occur in a sequence a large number of times. The radicals that are re-formed after each cycle are known as **chain carriers**.

The sequence of propagation reactions will continue until it is interrupted by a **termination reaction**. This involves the removal of the chain carrier, and often occurs by the recombination of radicals, as in the fourth step. For this reaction the sequence (137, 138) is *repeated about* 25 000 *times for each methyl radical formed initially*; this number is known as the **chain length**.

Because each methyl radical that is formed will go round the cycle (137, 138) so many times, and thereby produce many molecules of methane and carbon monoxide, these will be the *major products* of this reaction. The ethane, C_2H_6, that is formed in the termination step, and the $CHO\cdot$ (or its decomposition products) that is formed in the initiation step, will be *very minor* side products.†

Although the mechanism of any chain reaction may consist of many different kinds of elementary reaction, all chain reactions contain the three essential ingredients mentioned above:

1   initiation steps, which produce the chain carriers;

2   propagation steps, in which the chain carriers recycle and thus cause the conversion of reactants into products;

3   termination steps, which remove the recycling intermediates.

In Section 2.1.2 we said that for *non*-chain reactions the sum of the individual steps of a mechanism add up to the stoichiometric equation. As you saw above, for *chain* reactions it is the *addition of the propagation steps* that gives the overall stoichiometric equation.

* The fate of the $CHO\cdot$ formed in the initiation step is not specified in this mechanism. As you will see later, this turns out to be unimportant.

† If ethanal decomposes to give products other than carbon monoxide and methane, then not only is the stoichiometric equation inadequate, but also the rate of disappearance of ethanal will *not* be exactly equal to the rate of formation of methane and carbon monoxide. Thus, strictly speaking, J cannot be defined for this reaction. However, to simplify the analysis we shall continue to use J and the stoichiometric equation 133; the error so introduced will be minimal, since the concentration of side products is very low.

7.3 .The kinetics of a chain reaction

Having described the suggested mechanism for the decomposition of ethanal, our next step is to derive the chemical rate equation predicted by the proposed sequence and to confirm that it is consistent with the experimental rate equation.

■   Write down the theoretical rate equations for the reactant, CH_3CHO, the major products, CH_4 and CO, and the intermediates $CH_3\cdot$ and $CH_3CO\cdot$, based on the elementary steps of this mechanism.

■   The relevant rate equations are:

$$-\frac{d[CH_3CHO]}{dt} = k_1[CH_3CHO] + k_2[CH_3\cdot][CH_3CHO] \tag{140}$$

$$\frac{d[CH_4]}{dt} = k_2[CH_3\cdot][CH_3CHO] \tag{141}$$

$$\frac{d[CO]}{dt} = k_3[CH_3CO\cdot] \tag{142}$$

$$\frac{d[CH_3\cdot]}{dt} = k_1[CH_3CHO] - k_2[CH_3\cdot][CH_3CHO]$$
$$+ k_3[CH_3CO\cdot] - 2k_4[CH_3\cdot]^2 \tag{143}$$

$$\frac{d[CH_3CO\cdot]}{dt} = k_2[CH_3\cdot][CH_3CHO] - k_3[CH_3CO\cdot] \tag{144}$$

As they stand, it is not possible to simplify this set of theoretical rate equations because we do not know the concentrations of the intermediates $CH_3\cdot$ and $CH_3CO\cdot$. However, since this reaction exhibits time-independent stoichiometry, they can be simplified by using the steady-state approximation. In this case we shall apply it to *both chain carriers*, equations 143 and 144 becoming

$$0 = k_1[CH_3CHO] - \{k_2[CH_3\cdot][CH_3CHO] - k_3[CH_3CO\cdot]\} - 2k_4[CH_3\cdot]^2 \tag{145}$$

and $0 = k_2[CH_3\cdot][CH_3CHO] - k_3[CH_3CO\cdot]$ $\qquad\qquad$ (146)

respectively.

Comparing these two expressions reveals that the term in curly brackets in equation 145 is equal to zero, so this equation reduces to:

$$0 = k_1[CH_3CHO] - 2k_4[CH_3\cdot]^2 \tag{147}$$

Equation 147 gives us the steady-state concentration of $CH_3\cdot$, as:

$$[CH_3\cdot] = \left(\frac{k_1[CH_3CHO]}{2k_4}\right)^{1/2} \tag{148}$$

Substituting this expression for $[CH_3\cdot]$ into equation 141 (which is clearly the easiest to deal with) gives:

$$\frac{d[CH_4]}{dt} = k_2\left(\frac{k_1}{2k_4}\right)^{1/2}[CH_3CHO]^{3/2} \tag{149}$$

Since this reaction exhibits time-independent stoichiometry, we can write

$$J = -\frac{d[CH_3CHO]}{dt} = \frac{d[CH_4]}{dt} = \frac{d[CO]}{dt} = k_2\left(\frac{k_1}{2k_4}\right)^{1/2}[CH_3CHO]^{3/2} \tag{150}$$

Comparison of equations 150 and 134 shows that the mechanism *is* consistent with the experimental findings, and is therefore a very plausible candidate. There are, however, a number of experimental results that cannot be completely explained by this mechanism, principally that small traces of other products can be detected in this reaction (H_2, CH_3COCH_3, CH_3CH_2CHO and $CH_2{=}CH_2$). To account for these products, a mechanism involving 10 steps has been proposed. Nevertheless, the route to the major products is essentially the same as that given in equations 137 and 138.

■ What does equation 147 tell us about the rates of initiation and termination?

▨ Equation 147 tells us that when the steady state is set up, the rate of formation of $CH_3\cdot$ in the initiation step ($k_1[CH_3CHO]$) is equal to the rate of consumption of $CH_3\cdot$ in the termination step ($2k_4[CH_3\cdot]^2$).

Unfortunately, space does not allow anything more than a very limited introduction to chain reactions. As in this example, most chain reactions involve radical chain carriers, but cations and anions can also serve as reaction intermediates. In general, the concentration of the chain carrier remains low during the reaction, so the steady-state approximation is a particularly valuable tool for the analysis of such complex systems. Nevertheless, few examples can be analysed as simply as the one we have just examined. For instance, the seemingly straightforward reaction between hydrogen and bromine,

$$H_2(g) + Br_2(g) = 2HBr(g) \tag{151}$$

requires five steps to explain the experimental rate equation:

$$J = \frac{k_R[H_2][Br_2]^{1/2}}{1 + k_R'([HBr]/[Br_2])} \tag{152}$$

$$Br_2 \xrightarrow{\;k_1\;} Br\cdot + Br\cdot \qquad\qquad\qquad \textit{initiation} \tag{153}$$

$$Br\cdot + H_2 \xrightarrow{\;k_2\;} HBr + H\cdot \qquad\qquad\qquad \textit{propagation} \tag{154}$$

$$H\cdot + Br_2 \xrightarrow{\;k_3\;} HBr + Br\cdot \qquad\qquad\qquad \textit{propagation} \tag{155}$$

$$H\cdot + HBr \xrightarrow{\;k_{-2}\;} H_2 + Br\cdot \qquad\qquad\qquad \textit{inhibition} \tag{156}$$

$$Br\cdot + Br\cdot \xrightarrow{\;k_{-1}\;} Br_2 \qquad\qquad\qquad \textit{termination} \tag{157}$$

This mechanism differs from the one examined earlier in that it contains an extra step — an **inhibition step** — in which a hydrogen atom reacts with the product, HBr, to give a hydrogen molecule and a bromine atom. This is the reverse of equation 154.

The two examples of chain reactions we have discussed so far have quite complicated rate equations, but it should be stressed that this is not always the case. Thus, the dehydrogenation of ethane,

$$CH_3CH_3(g) = CH_2{=}CH_2(g) + H_2(g) \tag{158}$$

needs a five-step chain reaction to explain all of the experimental observations, yet it has a simple experimental rate equation of the form:

$$\frac{d[CH_2{=}CH_2]}{dt} = k_R[CH_3CH_3] \tag{159}$$

The decomposition of ethanal and the reaction between hydrogen and bromine are examples of **linear-chain reactions**; that is, when a radical takes part in the propagation steps, there is no net gain in the number of chain carriers. On the other hand, in a **branched-chain reaction** one active centre can react to produce more than one active centre. For example, the proposed mechanism for the reaction between hydrogen and oxygen,

$$2H_2(g) + O_2(g) = 2H_2O(g) \tag{160}$$

contains the following step:

$$H\cdot + O_2 \longrightarrow HO\cdot + O \tag{161}$$

This is known as a **chain-branching step**, since both HO· (the hydroxyl radical) and O can act as chain carriers; equations 162 and 163 show how these species react further:

$$O + H_2 \longrightarrow HO\cdot + H\cdot \tag{162}$$

$$HO\cdot + H_2 \longrightarrow H_2O + H\cdot \tag{163}$$

Equation 163 can be considered to be a propagation step because it involves the conversion of a reactant (H_2) to product (H_2O). The chain-branching steps shown in equations 161 and 162 can lead to a build up in the concentration of active centres, and thus a rapid rise in the rate of reaction. When the rate of reaction becomes catastrophically fast, this can often lead to explosions! Such an event occurs when hydrogen and oxygen are sparked together.

7.4 Summary of Section 7

1   The experimental rate equation for the thermal decomposition of ethanal can be accounted for by a chain reaction involving four steps, in which radicals are the chain carriers.

2   The radicals are formed in the first (initiation) step. The two propagation steps give rise to the products and continue the chain. The sequence will continue until interrupted by a termination reaction.

3   By applying the steady-state approximation to the chain carriers, it is possible to derive a chemical rate equation in terms of the concentrations of the reactants, which is in agreement with its experimental counterpart.

4   All chain reactions involve initiation, propagation and termination steps, and can usually be analysed using the steady-state approximation. However, most are more complex than the ethanal decomposition reaction, involving other steps such as inhibition or chain branching.

STUDY COMMENT   SAQs 11 and 12 give you the opportunity to analyse the mechanisms of chain reactions and to compare your results with experiment. Do try them.

SAQ 11   Reduce equation 140

$$-\frac{d[CH_3CHO]}{dt} = k_1[CH_3CHO] + k_2[CH_3\cdot][CH_3CHO] \tag{140}$$

into a form containing terms in ethanal (CH_3CHO) alone. If the experimental rate equation is

$$J = -\frac{d[CH_3CHO]}{dt} = \frac{d[CH_4]}{dt} = \frac{d[CO]}{dt} = k_R[CH_3CHO]^{3/2} \tag{164}$$

what does this say about the relative rates of initiation and product formation?

SAQ 12   The initial stages in the gas-phase thermal decomposition of monosilane, SiH_4, at temperatures in the range 650–700 K produce hydrogen and disilane, Si_2H_6, as major products:

$$2SiH_4(g) = Si_2H_6(g) + H_2(g) \tag{165}$$

The experimental rate equation for this reaction under these conditions takes the form:

$$J = k_R[SiH_4]^{3/2} \tag{166}$$

A radical chain mechanism has been proposed for this reaction as follows:

$$SiH_4 \xrightarrow{k_1} SiH_3\cdot + H\cdot \tag{167}$$

$$H\cdot + SiH_4 \xrightarrow{k_2} SiH_3\cdot + H_2 \tag{168}$$

$$SiH_3\cdot + SiH_4 \xrightarrow{k_3} Si_2H_6 + H\cdot \tag{169}$$

$$SiH_3\cdot + SiH_3\cdot \xrightarrow{k_4} Si_2H_6 \tag{170}$$

(a) Write down the theoretical rate equations for Si_2H_6, $H\cdot$ and $SiH_3\cdot$.

(b) By applying the steady-state approximation to both reactive intermediates, $H\cdot$ and $SiH_3\cdot$, and adding the resulting equations, obtain an expression for $[SiH_3\cdot]$ in terms of $[SiH_4]$.

(c) Derive the theoretical rate equation for Si_2H_6, and thus the chemical rate equation, in terms of the concentration of the reactant.

(d) Under what conditions is the chemical rate equation that you have derived in part (c) identical to the experimental rate equation?

OBJECTIVES FOR BLOCK 3

Now that you have completed Block 3, you should be able to do the following things:

1 Recognize valid definitions of, and use in a correct context, the terms, concepts and principles printed in bold type in the text and collected in the following Table.

List of scientific terms, concepts and principles used in Block 3

Term	Page no.
chain-branching step	39
chain carrier	37
chain length	37
chain reaction	36
chemical rate equation predicted by a mechanism	16
composite reaction mechanism	5
consecutive mechanism	10
inhibition step	39
initiation reaction	37
intermediate	7
linear- and branched-chain reactions	39
non-complementary electron transfer	24
opposing reactions	11
parallel reactions	11
pre-equilibrium	33
propagation reaction	37
radical	37
rate-limiting step	33
reactive intermediate	18
steady-state approximation	19
termination reaction	37
theoretical rate equation	16
time-dependent stoichiometry	6

2   Decide whether a reaction can proceed via a single elementary reaction or not, based on kinetic evidence or the detection of an intermediate. (SAQs 1 and 2)

3   Using the criteria outlined in Section 2, decide which of a selection of possi-bilities is (are) the most plausible mechanism(s). (SAQ 2)

4   Given the elementary steps of a particular mechanism, write down theoretical rate equations for reactants, intermediates and products. (SAQs 3, 5, 6, 7, 8 and 12)

5   Use the steady-state approximation to obtain an expression for the concentration of a reactive intermediate, and thus simplify a set of theoretical rate equations to a form that can be compared with the experimental rate equation. (SAQs 4, 6, 7, 8 and 12)

6   Use kinetic evidence to distinguish the most probable mechanism from a range of possibilities. (SAQ 7)

7   Discuss the consequences of particular terms dominating the chemical rate equation on (a) the chemical rate equation and (b) the proposed mechanism of a reaction. (SAQs 7 and 8)

8   Given the rate-limiting step of a mechanism, simplify the kinetics and thus derive a chemical rate equation. (SAQs 9 and 10)

9   Discuss the ideal features of a chain reaction, and by applying the steady-state approximation to the chain carriers, derive the chemical rate equation. (SAQs 11 and 12)

SAQ ANSWERS AND COMMENTS

SAQ 1 (Objective 2)

(a)   If this reaction were elementary, the stoichiometry would indicate a chemical rate equation of the form:

$$J = k_R[N_2O_5][NO] \tag{171}$$

Since the experimental rate equation does not involve the concentration of nitric oxide, NO, a composite reaction is likely to be involved.

(b)   The chemical rate equation obtained from the stoichiometry, assuming that the reaction is elementary, is identical with that determined experimentally. Thus, kinetic data do *not* indicate that this reaction is composite. Indeed, this was one of the reasons why the reaction was thought for so long to be elementary. This situation is frequently encountered in mechanistic studies. A number of observations may be consistent with a reaction proceeding in an elementary fashion, and the composite nature of the reaction is only revealed when other mechanistic tests are applied.

SAQ 2 (Objectives 2 and 3)

(a)   The stoichiometric equation involves four reactant species. If this reaction were elementary, it would involve a four-centre collision! This is extremely unlikely, and so we conclude that the reaction is composite. Secondly, if this reaction were elementary, the rate of change of $[V^{2+}]$ and $[V^{3+}]$ should be related as follows:

$$J = -\frac{d[V^{2+}]}{dt} = \frac{1}{2}\frac{d[V^{3+}]}{dt} \tag{172}$$

We would therefore expect the concentration of V^{3+} to change *twice as fast* as the concentration of V^{2+} changes. But the opposite behaviour is observed at the beginning of the reaction. This is similar to the situation in the hydrogen peroxide/nitrous acid reaction, and is another example of time-dependent stoichiometry. This provides further support for a composite reaction.

(b)   To choose the most plausible mechanism, we must apply the criteria given in stage 2 of Box 1. The question allows us to assume that the chemistry in all three mechanisms is reasonable.

(i)   This mechanism can be discounted because it does not add up to the stoichiometric equation. In fact, it gives

$$V^{2+}(aq) + VO^{2+}(aq) + 2H^+(aq) = V^{3+}(aq) + VO^{3+}(aq) + H_2(g) \qquad (173)$$

(ii) This mechanism is the most plausible. It does not disagree with any of the criteria we employ for selecting a mechanism.

(iii) This mechanism is very unlikely because the third step involves a four-centre collision.

SAQ 3 (Objective 4)

First we give each step a rate constant

$$2ClO^- \xrightarrow{\ k_1\ } ClO_2^- + Cl^- \qquad (174)$$

$$ClO_2^- + ClO^- \xrightarrow{\ k_2\ } ClO_3^- + Cl^- \qquad (175)$$

Next we write down the theoretical rate equations from the stoichiometry of each elementary step.

(a) The hypochlorite ion, ClO^-

This is *consumed* in both steps; considering the rate of change in concentration for each step alone:

$$-\frac{1}{2}\left(\frac{d[ClO^-]}{dt}\right)_1 = k_1[ClO^-]^2 \qquad (176)$$

two molecules have to collide! remember? do have to collide. $k \cdot [ClO^-][ClO^-]$

$$-\left(\frac{d[ClO^-]}{dt}\right)_2 = k_2[ClO_2^-][ClO^-] \qquad (177)$$

Note the factor of $\frac{1}{2}$ that appears in equation 176. This is a result of our definition of J for the first elementary step.

The overall change in concentration of hypochlorite ion is given by

$$\frac{d[ClO^-]}{dt} = \left(\frac{d[ClO^-]}{dt}\right)_1 + \left(\frac{d[ClO^-]}{dt}\right)_2 \qquad (178)$$

Thus,

$$\frac{d[ClO^-]}{dt} = -2k_1[ClO^-]^2 - k_2[ClO_2^-][ClO^-] \qquad (179)$$

(b) The chloride ion, Cl^-

This is formed in both steps:

$$\left(\frac{d[Cl^-]}{dt}\right)_1 = k_1[ClO^-]^2 \qquad (180)$$

$$\left(\frac{d[Cl^-]}{dt}\right)_2 = k_2[ClO_2^-][ClO^-] \qquad (181)$$

Thus,

$$\frac{d[Cl^-]}{dt} = k_1[ClO^-]^2 + k_2[ClO_2^-][ClO^-] \qquad (182)$$

(c) The chlorite ion, ClO_2^-

This intermediate is formed in the first step and consumed in the second:

$$\left(\frac{d[ClO_2^-]}{dt}\right)_1 = k_1[ClO^-]^2 \qquad (183)$$

$$-\left(\frac{d[ClO_2^-]}{dt}\right)_2 = k_2[ClO_2^-][ClO^-] \qquad (184)$$

Thus,

$$\frac{d[ClO_2^-]}{dt} = k_1[ClO^-]^2 - k_2[ClO_2^-][ClO^-] \qquad (185)$$

(d) The chlorate ion, ClO_3^-

This is formed in the second step alone. Thus,

$$\frac{d[ClO_3^-]}{dt} = k_2[ClO_2^-][ClO^-] \qquad (186)$$

With practice, you should be able to develop the skill of writing down the theoretical rate equation for a particular species in a reaction mechanism without the need to consider each step separately.

SAQ 4 (Objective 5)

Since this reaction exhibits time-independent stoichiometry, the concentration of the intermediate will be very low during the reaction and we can apply the steady-state approximation. This involves assuming

$$\frac{d[ClO_2^-]}{dt} = 0 \qquad (187)$$

Thus, equation 185 becomes

$$0 = k_1[ClO^-]^2 - k_2[ClO_2^-][ClO^-] \qquad (188)$$

Rearranging equation 188 gives us an expression for $[ClO_2^-]$:

$$[ClO_2^-] = \frac{k_1[ClO^-]^2}{k_2[ClO^-]} = \frac{k_1[ClO^-]}{k_2} \qquad (189)$$

Substituting this into equations 179, 182 and 186 gives

$$\frac{d[ClO^-]}{dt} = -2k_1[ClO^-]^2 - k_2 \frac{k_1}{k_2}[ClO^-][ClO^-] = -3k_1[ClO^-]^2 \qquad (190)$$

$$\frac{d[Cl^-]}{dt} = k_1[ClO^-]^2 + k_2 \frac{k_1}{k_2}[ClO^-][ClO^-] = 2k_1[ClO^-]^2 \qquad (191)$$

$$\frac{d[ClO_3^-]}{dt} = k_2 \frac{k_1}{k_2}[ClO^-][ClO^-] = k_1[ClO^-]^2 \qquad (192)$$

Combining equations 190 and 191 with 192 gives the chemical rate equation predicted by this mechanism:

$$J = -\frac{1}{3}\frac{d[ClO^-]}{dt} = \frac{1}{2}\frac{d[Cl^-]}{dt} = \frac{d[ClO_3^-]}{dt} = k_1[ClO^-]^2 \qquad (193)$$

So any of the three theoretical rate equations (179, 182 or 186) could be used to derive the chemical rate equation.

The experimental rate equation is of the form

$$J = -\frac{1}{3}\frac{d[ClO^-]}{dt} = \frac{1}{2}\frac{d[Cl^-]}{dt} = \frac{d[ClO_3^-]}{dt} = k_R[ClO^-]^2 \qquad (194)$$

Equations 193 and 194 are both second order in ClO^-, and are identical if $k_R = k_1$.

SAQ 5 (Objective 4)

(a) Starting with the reactant, A: this is consumed in the forward reaction but formed in the reverse reaction.

Treating each step in isolation, we can write

$$\left(\frac{d[A]}{dt}\right)_1 = -k_1[A] \tag{195}$$

$$\left(\frac{d[A]}{dt}\right)_{-1} = k_{-1}[B] \tag{196}$$

Note that although the reverse reaction is written from right to left, the reactants and the products of a particular elementary step are defined by the direction of the arrow.

The overall rate equation for A is obtained by summing equations 195 and 196:

$$\frac{d[A]}{dt} = -k_1[A] + k_{-1}[B] \tag{197}$$

By similar reasoning,

$$\frac{d[B]}{dt} = k_1[A] - k_{-1}[B] \tag{198}$$

(b) At equilibrium there is no net change in the concentration of A and B, so d[A]/dt and d[B]/dt are both zero.

(c) At equilibrium both equations 197 and 198 become

$$0 = k_1[A]_{eq} - k_{-1}[B]_{eq} \tag{199}$$

(In other words, at equilibrium, the rates of the forward and reverse reactions are the same.) Thus

$$\frac{k_1}{k_{-1}} = \frac{[B]_{eq}}{[A]_{eq}} \tag{200}$$

The ratio on the right-hand side of equation 200 corresponds to the expression for the equilibrium constant K_c for the reaction:

$$\frac{k_1}{k_{-1}} = \frac{[B]_{eq}}{[A]_{eq}} = K_c \tag{201}$$

This is a perfectly general result, namely that *for opposing elementary steps, the equilibrium constant is the ratio of the individual rate constants for the forward and reverse reactions.*

SAQ 6 (Objectives 4 and 5)

(a) (i) Tl^I is formed in the second step (equation 96) only. Thus,

$$\frac{d[Tl^I]}{dt} = k_2[\mathbf{Fe^I}][Tl^{III}] \tag{202}$$

(ii) $\mathbf{Fe^I}$ is formed in the forward reaction of equation 95, and consumed in the reverse reaction of equation 95 and in the second step (equation 96). Thus,

$$\frac{d[\mathbf{Fe^I}]}{dt} = k_1[Fe^{II}]^2 - k_{-1}[Fe^{III}][\mathbf{Fe^I}] - k_2[\mathbf{Fe^I}][Tl^{III}]$$

$$= k_1[Fe^{II}]^2 - \{k_{-1}[Fe^{III}] + k_2[Tl^{III}]\}[\mathbf{Fe^I}] \tag{203}$$

(b) Assuming that d[$\mathbf{Fe^I}$]/dt is negligible in comparison with the other terms in equation 203, we can apply the steady-state approximation:

$$\frac{d[\mathbf{Fe^I}]}{dt} = 0 = k_1[Fe^{II}]^2 - \{k_{-1}[Fe^{III}] + k_2[Tl^{III}]\}[\mathbf{Fe^I}] \tag{204}$$

so $$[\mathbf{Fe^I}] = \frac{k_1[Fe^{II}]^2}{k_{-1}[Fe^{III}] + k_2[Tl^{III}]} \tag{205}$$

(c)   Combining this with equation 202 gives

$$\frac{d[Tl^I]}{dt} = \frac{k_1 k_2 [Tl^{III}][Fe^{II}]^2}{k_{-1}[Fe^{III}] + k_2[Tl^{III}]} \tag{206}$$

Defining J from the overall stoichiometric equation (equation 80 in Section 5.1),

$$J = \frac{d[Tl^I]}{dt} = \frac{k_1 k_2 [Tl^{III}][Fe^{II}]^2}{k_{-1}[Fe^{III}] + k_2[Tl^{III}]} \tag{207}$$

SAQ 7 (Objectives 4, 5, 6 and 7)

(a)   The two theoretical rate equations are:

(i) $$\frac{d[NO_3]}{dt} = k_1[NO][O_2] - k_{-1}[NO_3] - k_2[NO_3][NO] \tag{208}$$

(ii) $$\frac{1}{2}\frac{d[NO_2]}{dt} = k_2[NO_3][NO] \tag{209}$$

Applying the steady-state approximation to NO_3, we can assume

$$\frac{d[NO_3]}{dt} = 0 = k_1[NO][O_2] - \{k_{-1} + k_2[NO]\}[NO_3] \tag{210}$$

Rearrangement gives us an expression for $[NO_3]$:

$$[NO_3] = \frac{k_1[NO][O_2]}{k_{-1} + k_2[NO]} \tag{211}$$

Substituting this into equation 209 gives

$$\frac{1}{2}\frac{d[NO_2]}{dt} = \frac{k_1 k_2 [NO]^2[O_2]}{k_{-1} + k_2[NO]} \tag{212}$$

Thus, from the definition of J for this reaction, we can write

$$J = \frac{1}{2}\frac{d[NO_2]}{dt} = \frac{k_1 k_2 [NO]^2[O_2]}{k_{-1} + k_2[NO]} \tag{213}$$

If k_{-1} is much greater than $k_2[NO]$, it will dominate the denominator and the expression reduces to

$$J = \left(\frac{k_1 k_2}{k_{-1}}\right)[NO]^2[O_2] \tag{214}$$

which is in the same form as the experimental rate equation ($k_R = k_1 k_2/k_{-1}$).

(b)   The theoretical rate equation for NO will contain three terms, since NO is involved in both the forward and reverse reactions of step 1, and also in step 2:

$$\frac{d[NO]}{dt} = -k_1[NO][O_2] + k_{-1}[NO_3] - k_2[NO_3][NO] \tag{215}$$

Since substituting for $[NO_3]$ is required in two of the terms, the process of simplification will be more involved than it was for equation 209.

The theoretical rate equation for O_2 will contain two terms:

$$\frac{d[O_2]}{dt} = -k_1[NO][O_2] + k_{-1}[NO_3] \tag{216}$$

Substitution for $[NO_3]$ is required in one term, so the process of simplification will be only marginally more involved than it was for equation 209.

(c)   If this reaction proceeds by a single elementary step (equation 120), the chemical rate equation will be

$$J = k_1'[NO]^2[O_2] \tag{217}$$

This is also compatible with the experimental rate equation, and thus we *cannot* use kinetic data to distinguish between these two possible mechanisms.

SAQ 8 (Objectives 4, 5 and 7)

The theoretical rate equations based on this mechanism are:

$$-\frac{d[A]}{dt} = k_1[A] \tag{218}$$

$$\frac{d[B]}{dt} = k_1[A] - k_2[B] \tag{219}$$

$$\frac{d[C]}{dt} = k_2[B] \tag{220}$$

By applying the steady-state approximation to B, equation 219 can be put equal to zero, which gives the following expression for [B]:

$$[B] = \frac{k_1}{k_2}[A] \tag{221}$$

Substituting this expression for [B] into equation 220 gives

$$\frac{d[C]}{dt} = k_1[A] \tag{222}$$

Thus,

$$J = -\frac{d[A]}{dt} = \frac{d[C]}{dt} = k_1[A] \tag{223}$$

This agrees with the conclusion we came to in the AV sequence. The above analysis shows that when $k_2 \gg k_1$, the chemical rate equation contains the smaller rate constant, k_1, alone. In other words, *the first step is the rate-limiting step*. This conclusion is the same, irrespective of whether we examine the differential forms (as here) or the integrated rate equations (as in the AV sequence).

SAQ 9 (Objective 8)

If the second step is rate limiting, J for the overall reaction can be set equal to the rate equation for this step:

$$J = k_2[HClO][I^-] \tag{224}$$

Assuming that the first step is a rapidly established pre-equilibrium, the concentration of the intermediate, HClO, will be governed by the size of the equilibrium constant for this step:

$$K_c = \frac{k_1}{k_{-1}} = \frac{[HClO][OH^-]}{[ClO^-][H_2O]} \tag{225}$$

Rearranging

$$[HClO] = \frac{k_1}{k_{-1}}\frac{[ClO^-][H_2O]}{[OH^-]} \tag{226}$$

Substituting this expression for [HClO] into equation 224 gives

$$J = \frac{k_1 k_2}{k_{-1}}\frac{[ClO^-][H_2O][I^-]}{[OH^-]} \tag{227}$$

Since the concentration of water in aqueous solution can be assumed to be constant, the chemical rate equation predicted by this mechanism is identical to the experimental rate equation (127), where

$$k_R = \frac{k_1 k_2}{k_{-1}}[H_2O] \tag{228}$$

Notice that in this case the experimental rate constant is a function of three elementary rate constants.

SAQ 10 (Objective 8)

In this question, both pre-equilibria are used to 'remove' the intermediate from the chemical rate equation. Using the strategy outlined in Box 5, and assuming that the last step is rate limiting,

$$J = k_3[OH^-][HIO] \tag{229}$$

The concentration of the intermediate HIO can be obtained from inspection of the second pre-equilibrium (equation 131):

$$[HIO] = \frac{k_2}{k_{-2}} \frac{[HClO][I^-]}{[Cl^-]} \tag{230}$$

Thus, combining equations 229 and 230 gives

$$J = \left(\frac{k_2 k_3}{k_{-2}}\right) \frac{[OH^-][HClO][I^-]}{[Cl^-]} \tag{231}$$

The concentration of the intermediate HClO can be obtained from inspection of the first pre-equilibrium (equation 128):

$$[HClO] = \frac{k_1}{k_{-1}} \frac{[ClO^-][H_2O]}{[OH^-]} \tag{232}$$

Combining equations 231 and 232 gives

$$J = \frac{k_1 k_2 k_3}{k_{-1} k_{-2}} \frac{[ClO^-][H_2O][I^-]}{[Cl^-]} \tag{233}$$

This chemical rate equation is in the form required in the question, but is *not* consistent with the experimental rate equation (127). Hence we conclude that the assumptions made in the question are *not* valid.

SAQ 11 (Objective 9)

Substituting for $[CH_3\cdot]$ from equation 148 in equation 140, we get

$$-\frac{d[CH_3CHO]}{dt} = k_1[CH_3CHO] + k_2\left(\frac{k_1}{2k_4}\right)^{1/2}[CH_3CHO]^{3/2} \tag{234}$$

Clearly, this expression reduces to a form compatible with the experimental rate equation if the second term is very much larger than the first. If this is the case, the rate of product formation (the second term in equation 234) is bigger than the rate of initiation (the first term in equation 234). This must indeed be the case, for otherwise CH_4 and CO would not be the major products.

SAQ 12 (Objectives 4, 5 and 9)

(a)
$$\frac{d[Si_2H_6]}{dt} = k_3[SiH_3\cdot][SiH_4] + k_4[SiH_3\cdot]^2 \tag{235}$$

$$\frac{d[H\cdot]}{dt} = k_1[SiH_4] - k_2[H\cdot][SiH_4] + k_3[SiH_3\cdot][SiH_4] \tag{236}$$

$$\frac{d[SiH_3\cdot]}{dt} = k_1[SiH_4] + k_2[H\cdot][SiH_4] - k_3[SiH_3\cdot][SiH_4] - 2k_4[SiH_3\cdot]^2 \tag{237}$$

(b)   Applying the steady-state approximation to $H\cdot$ and $SiH_3\cdot$ gives:

$$0 = k_1[SiH_4] - k_2[H\cdot][SiH_4] + k_3[SiH_3\cdot][SiH_4] \tag{238}$$

$$0 = k_1[SiH_4] + k_2[H\cdot][SiH_4] - k_3[SiH_3\cdot][SiH_4] - 2k_4[SiH_3\cdot]^2 \tag{239}$$

Adding these equations gives:

$$0 = 2k_1[SiH_4] - 2k_4[SiH_3\cdot]^2 \tag{240}$$

so

$$[SiH_3\cdot] = \left(\frac{k_1[SiH_4]}{k_4} \right)^{1/2} \tag{241}$$

(c) Substituting for $[SiH_3\cdot]$ in the expression for $d[Si_2H_6]/dt$ (equation 235),

$$J = \frac{d[Si_2H_6]}{dt} = k_3 \left(\frac{k_1}{k_4} \right)^{1/2} [SiH_4]^{3/2} + k_1[SiH_4] \tag{242}$$

(d) As in SAQ 11, this expression is identical with the experimental rate equation if the rate of initiation, $k_1[SiH_4]$, is much less than that of product formation via the propagation steps, $k_3(k_1/k_4)^{1/2}[SiH_4]^{3/2}$.

PHYSICAL CHEMISTRY

PRINCIPLES OF CHEMICAL CHANGE

BLOCK 4
HOMOGENEOUS CATALYSIS

CONTENTS

1 INTRODUCTION

The reaction of hydrogen and oxygen to give water,

$$H_2(g) + \tfrac{1}{2}O_2(g) = H_2O(g) \tag{1}$$

has a $\Delta G_m^{\ominus}$ value of $-228.5\,\text{kJ mol}^{-1}$ at 298.15 K; that is, $K^{\ominus} = 1.08 \times 10^{40}$. This indicates that equilibrium in this system lies well over to the right-hand side. However, if gaseous hydrogen and oxygen are carefully mixed in a pure state, nothing happens; although this reaction is thermodynamically favourable, its *rate* is effectively zero. But, when some finely divided platinum is added to the mixture, the reaction takes place so rapidly that the platinum is heated to incandescence and an explosion generally occurs (you will recall from Block 3, Section 7.3 that this is an example of a branched chain reaction). In this system the platinum has acted as a *catalyst,* a substance that speeds up the reaction without being consumed itself.

You have already come across a reference to the practical importance of catalysis. In the exercise at the end of Block 1 you examined the factors that affect the yield of methanol obtained from the following process:

$$CO(g) + 2H_2(g) = CH_3OH(g) \tag{2}$$

■ The value of $\Delta H_m^{\ominus}$ (298.15 K) for this process was found to be $-90.2\,\text{kJ mol}^{-1}$. How would you increase the *equilibrium yield* of methanol?

▨ The equilibrium yield of methanol could be increased by keeping the temperature low and by increasing the overall pressure.

■ What other factors influence the *actual yield* of methanol obtained from this process?

▨ When dealing with the actual yield rather than the equilibrium yield, we should also take into account the factors that influence the rate of reaction — for example, the temperature or the use of a catalyst. Secondly, synthesis gas, CO and H_2, can undergo reactions to produce a whole range of compounds — such as methane and other alkanes, ethane-1,2-diol ($HOCH_2CH_2OH$) and other alcohols. Furthermore, all these reactions are thermodynamically favourable. So we must also consider the **selectivity** of this reaction; that is, how much of the starting material is converted into the desired product, as opposed to the amount that is lost in unwanted side reactions?

■ What are the disadvantages of operating this process at a higher temperature?

▨ Although raising the temperature would increase the rate of reaction, it would also result in a lower equilibrium yield because this is an exothermic reaction. Under these circumstances, the attainment of a reasonable equilibrium yield may involve difficult and expensive operating conditions, such as the use of very high pressures.

The much more effective way of improving the yield is to employ a catalyst. If the *right* catalyst is chosen, not only will the equilibrium yield be obtained in a shorter time, but the conversion into methanol will be the *only* reaction that is promoted, so the selectivity will also improve. In fact, for this reaction, a copper-based catalyst is used, which is highly selective for methanol.

Catalysts are used to accelerate an ever-increasing array of chemical reactions. They provide the key to the greatly expanded modern chemical and petroleum industries: about 90% of chemical manufacturing processes are catalytic. Sulfuric acid, nitric acid, ammonia, edible oils, synthetic rubber, plastics and a whole range of organic chemicals, are now produced almost completely by catalytic processes. Petroleum refining, which provides the largest volume of industrial products, consists almost entirely of a series of catalytic reactions. Catalysts are used to ease our pollution problems, reducing noxious emissions from cars and chemical plants, and eliminating odours. Above all, catalysts are essential to life itself: *enzymes* function as catalysts that regulate the chemical reactions on which life depends.

The earliest applications of catalysis were in the production of wine, vinegar and soap, processes that date back to antiquity. However, the first real industrial application came in 1746, when nitric oxide (nitrogen monoxide) was used to accelerate the oxidation of sulfur dioxide to sulfur trioxide:

$$SO_2(g) + \tfrac{1}{2}O_2(g) \xrightarrow{\text{NO}} SO_3(g) \tag{3}$$

This reaction formed part of the manufacture of sulfuric acid by a method known as the 'lead chamber process'. Catalytic production of inorganic compounds rapidly increased during the nineteenth and early twentieth centuries as new catalytic processes were discovered. Even established processes were improved. For example, in 1831, Peregrine Phillips, a vinegar chemist from Bristol, developed the use of platinum metal as a replacement for nitric oxide in the production of sulfuric acid. This system became known as the 'contact process'.

The use of catalysts to produce organic compounds first started towards the end of the nineteenth century, with the studies of the eminent French chemist Paul Sabatier on the reaction between hydrogen and unsaturated compounds. This led to the hydrogenation of unsaturated fats for food, using a nickel catalyst. During the twentieth century the search for new catalysts has continued unabated, with the production of hydrocarbon fuels growing into the largest by far of the industrial catalytic applications.

Space has allowed only the briefest sketch of the role of catalysis in chemical processes, but none the less this should be sufficient to impress on you its importance. The advances made in our understanding of catalysis and its applications in the past 100 years or so have been remarkable, and it is essential that this progress be maintained. We must continue to conserve energy by using raw materials in an effective manner, and by developing even more efficient processes. To this end, recent advances in biotechnology have generated a wave of excitement about the prospective application of nature's catalysts, in the form of enzymes and micro-organisms, to produce food, drink, pharmaceuticals and industrial chemicals.

It is because catalysis is of such vital importance that we shall examine the underlying principles of catalytic action in detail in the next few Blocks. For practical reasons, the subject area is divided into two parts. First, we concentrate in this Block on **homogeneous catalysis**, in which the catalysts and the reacting substances are present together in a single state of matter, usually a gas or a liquid. The acceleration of the oxidation of sulfur dioxide to sulfur trioxide by nitric oxide is an example of homogeneous catalysis, since all are present as a homogeneous mixture in the gas phase.

In Block 5 we shall examine **heterogeneous catalysis**, in which the catalyst is in a different phase from that of the reactants — most commonly, gaseous or liquid reactants over a solid catalyst. The use of platinum metal to catalyse the formation of sulfur trioxide is an example of heterogeneous catalysis. In this reaction, the catalyst is a solid and the reactants are gases, the catalytic reaction taking place *at the surface* of the platinum. At present most industrial processes involve heterogeneous catalysis.

The basic principles of heterogeneous and homogeneous catalysis are very similar. As you will see, the main differences are in the appearance and properties of the catalysts and in their practical industrial applications. There are also differences in the experimental techniques used to investigate the two types of catalytic process.

Block 4 itself deals with homogeneous catalysis, and is divided into three parts. First, we shall examine the theoretical aspects of catalysis. Because we are dealing with the *rates* of chemical reactions, our knowledge of kinetics helps us to establish that the catalyst provides an easier pathway for reaction. Next, we review the applications of homogeneous catalysis to industrial processing, using transition metal catalysts as an example. Thirdly, we look at the kinetic and mechanistic aspects of enzyme reactions, and discover that the behaviour of these complex biological catalysts can be adequately explained using relatively simple models.

2 CATALYSIS:
THE GENERAL PRINCIPLES

In this Section we shall discuss the general principles of catalysis. Most of the examples will involve homogeneous systems, but you should always bear in mind that the arguments and definitions presented here require only slight modification to be equally useful in heterogeneous catalysis.

2.1 Catalysis — a definition

The term 'catalysis' was first used by the great Swedish chemist Jöns Jacob Berzelius in 1835. He was the first scientist to try to make sense of this phenomenon. He realized that the catalyst speeded up the reaction without being consumed, but he could only describe the mode of action in terms of some mysterious force. The pioneering work of L. Wilhelmy on the hydrolysis of methyl ethanoate (methyl acetate, CH_3COOCH_3) put the study of catalysis on a more quantitative footing. He was the first to appreciate differences in the *rates* of chemical reaction, which gets to the heart of catalysis.

The next milestone occurred in 1877, when Georges Lemoine demonstrated that the decomposition of hydrogen iodide to hydrogen and iodine reached the *same* equilibrium composition at 350 °C, *irrespective of whether the reaction was carried out rapidly in the presence of a platinum catalyst, or slowly on its own in the gas phase*. This led to Wilhelm Ostwald redefining catalysts as substances that changed the rate of a given reaction *without* modification of the 'energy factors' of the reaction.

■ Why do you think the presence of a catalyst does not affect the Gibbs free energy change (that is, the value of $\Delta G_m^\ominus$) and hence the equilibrium position of a reaction?

■ Because the free energy change of a reaction is dependent only on the *initial* and *final* states of the system, it is independent of the actual pathway whereby that change is effected. Equally, because the catalyst is *not consumed* during the reaction, the free energy change in the presence of the catalyst will be the same as that in its absence.

This leads us to the important conclusion that *a catalyst cannot change a thermodynamically unfavourable reaction into a favourable one*. In fact, the definition of a **catalyst** has changed little since Ostwald's day: in this Course we shall define it as follows.

A catalyst is a substance that increases the rate of a chemical reaction without itself being consumed or altering the position of thermodynamic equilibrium.

This statement can be amplified:

- As you will see later, although a catalyst is not consumed, as such, it *does* take part in the reaction.

- The *physical* state of a solid catalyst (see Block 5) may change during its use.

■   If a catalyst increases the rate of the forward reaction *without changing the position of equilibrium*, what effect does the catalyst have on the rate of the reverse reaction? [*Hint*   Consider the rates of the forward and back reactions at equilibrium.]

▨   At equilibrium, the rates of the forward and reverse reactions are the same. Thus, if the rate of the forward reaction is increased, and the position of equilibrium is not to alter, the rate of the reverse reaction must also increase by a similar proportion.

2.2   Catalysis — the mode of action

Thermodynamics can tell us a lot about the equilibrium state of a system but nothing about the speed at which that state is achieved. Catalysts have *no* effect on the position of equilibrium, but they do have a *dramatic* effect on the rate at which it is attained. Catalysis, therefore, lies in the realm of kinetics, and much information can be obtained from the analysis of the rate equations of catalysed and uncatalysed reactions.

A simple, but important, example of homogeneous catalysis is involved in the decomposition of hydrogen peroxide:

$$2H_2O_2(aq) = 2H_2O(l) + O_2(g) \tag{4}$$

The reaction has many applications, notably in bleaching (straw hats and tripe, for example). The uncatalysed reaction is very slow, having the following experimental rate equation:

$$J = k_R[H_2O_2]^2 \tag{5}$$

When a small amount of an iodide salt is added to the reaction mixture, the same net reaction occurs, but it goes a lot faster. The experimental rate equation takes the form:

$$J = k_R'[H_2O_2][I^-] \tag{6}$$

■   Is this an example of catalysis, and, if so, what is the catalyst? How will this affect the equilibrium yield of oxygen?

▨   This is an example of a catalysed reaction; in the presence of iodide ion the reaction goes a lot faster. In this case, this is reflected in the experimental rate equation (equation 6), which now contains a term in the concentration of iodide ion, I^-. However, the stoichiometry of the overall reaction, given by equation 4, shows that iodide ions are not consumed in this reaction. Thus, the presence of I^- cannot alter the final position of equilibrium. Given sufficient time, the uncatalysed reaction would give the same equilibrium yield of reactants and products as the catalysed reaction.

■   For the decomposition of hydrogen peroxide, the uncatalysed and catalysed reactions have the experimental rate equations 5 and 6, respectively. As these rate equations differ, what can we say about the mechanisms of these two processes?

▨   As you saw in Block 3, the experimental rate equation reflects the mechanism of a reaction. Quite often, different mechanisms give rise to different chemical rate equations, and this is one way of weeding out unsuitable pathways (as in Block 3, Section 5.1). The corollary is that different experimental rate equations usually arise from different mechanisms.

The conclusion — that catalysed and uncatalysed reactions proceed via different mechanisms — is at the heart of catalysis, both homogeneous and heterogeneous. The *purpose* of using a catalyst is to provide a *faster* pathway from the reactants to the products.

■ From a consideration of the experimental rate equation 6 and the stoichiometric equation 4, do you think that the catalysed reaction involves a composite mechanism or not?

■ If this reaction were elementary, the stoichiometry predicts a chemical rate equation of the form

$$J = k_R[H_2O_2]^2 \tag{7}$$

which is at odds with that obtained experimentally. Even if we were to include the catalyst in the elementary reaction,

$$2H_2O_2 + I^- \longrightarrow 2H_2O + O_2 + I^- \tag{8}$$

the predicted chemical rate equation,

$$J = k_R[H_2O_2]^2[I^-] \tag{9}$$

would still be different from equation 6. Clearly the reaction is composite.

As you can prove for yourself by doing SAQ 1, one mechanism for the catalysed route which *is* consistent with the experimental rate equation 6 goes as follows:

$$H_2O_2 + I^- \xrightarrow{k_1} H_2O + IO^- \tag{10}$$

$$IO^- + H_2O_2 \xrightarrow{k_2} H_2O + O_2 + I^- \tag{11}$$

SAQ 1 (revision) Assuming that the intermediate IO^- is a reactive intermediate, use the steady-state approximation to obtain the chemical rate equation predicted by this mechanism. Is it consistent with the experimental rate equation 6?

Notice the intimate role of the catalyst in this mechanism — a role that is certainly *not* apparent from the stoichiometric equation. The catalyst actually undergoes chemical reactions with the reactants, involving the making and breaking of bonds, and the formation of an intermediate containing the catalyst. It is because of this that the catalyst appears in the rate equation. However, although it is consumed in the first step of the mechanism, it is regenerated in the second step, so its concentration is the same at the end as at the beginning of the reaction (and hence it does *not* appear in the stoichiometric equation), in accord with the definition of a catalyst.

Having determined that the catalyst provides a speedier pathway from reactants to products, we now turn our attention to establishing *why* this alternative route is quicker.

■ Generally, the concentration of the catalyst is much less than that of the reactants; only a relatively small amount of iodide ion is required to catalyse the decomposition of hydrogen peroxide. This being the case, what can you deduce about the relative sizes of k_u and k_c from a comparison of the experimental rate equations of the uncatalysed (equation 12) and catalysed (equation 13) reactions?

$$J_u = k_u[H_2O_2]^2 \tag{12}$$

$$J_c = k_c[H_2O_2][I^-] \tag{13}$$

(Remember that the rate of the catalysed reaction, J_c, is much greater than that of the uncatalysed process, J_u).

■ If the concentration of iodide ion is less than that of hydrogen peroxide, then the term $[H_2O_2]^2$ must be larger than the term $[H_2O_2][I^-]$. If this is the case, the only way for J_c to be greater than J_u is for the rate constant for the catalysed process, k_c, to be much greater than that for the uncatalysed reaction, k_u.

Now, as you know from Block 2, the size of experimental rate constants, such as k_u and k_c, depends on two factors — the energy of activation, E_a, and the A-factor, as reflected by the Arrhenius equation,

$$k_R = A \exp(-E_a/RT) \tag{14}$$

Table 1 lists data for the A-factors and energies of activation for this reaction with various catalysts. Notice that in each case the rate constant for the catalysed route is much larger than that for the uncatalysed process.

Table 1   Comparison of kinetic parameters for the uncatalysed and catalysed decomposition of aqueous hydrogen peroxide

Catalyst	Experimental rate equation	Temperature/°C	k_R/dm^3 mol^{-1} s^{-1}	A/dm^3 mol^{-1} s^{-1}	E_a/kJ mol^{-1}
none	$J = k_R[H_2O_2]^2$	22	$\sim 10^{-7}$	$\sim 10^6$	70–75
I$^-$	$J = k_R[H_2O_2][I^-]$	25	1×10^{-2}	1×10^8	57
Fe^{2+}	$J = k_R[H_2O_2][Fe^{2+}]$	22	56.0	2×10^9	42
catalase (enzyme)	$J = k_R[H_2O_2][catalase]$	22	3.5×10^7	6×10^8	7

■   How do the values of the A-factor and the energy of activation vary between the uncatalysed and catalysed reactions? Which of these two factors contributes more to the larger rate constants of the catalysed processes?

▨   In each case, the A-factor increases and the activation energy decreases on going from the uncatalysed to the catalysed pathway. *Both* of these changes will result in larger rate constants. However, it is the energy of activation that contributes more to the increase in the value of k_R. Although the change in activation energy appears to be of a smaller magnitude than that of the A-factor, such changes can lead to large variations in k_R, because the rate constant depends on the term $\exp(-E_a/RT)$. For example, even if there were no change in the A-factor, a decrease in activation energy by about a half from 75 to 42 would result in an increase in the rate constant by a factor of about 7×10^5!

We deduce that the increase in the rate constant on going from an uncatalysed to a catalysed reaction is largely due to the decrease in the energy of activation, in this case contributing to a change in k_R by a factor of up to 10^{14}. The value of A also increases, but by a relatively small amount, a factor of less than 10^3.

We can summarize our conclusions on the mode of action of a catalyst as follows:

A catalyst increases the rate of reaction in both the forward and reverse directions, usually by providing an alternative mechanism involving a lower activation energy.

STUDY COMMENT   You should now attempt the following SAQ. It introduces you to two particularly important types of catalyst.

SAQ 2   Ethane-1,2-diol, **1**, is produced on an industrial scale by the aqueous hydrolysis of oxirane (C_2H_4O), **2**, as follows:

$$H_2O(l) + H_2C\overset{O}{\overbrace{\quad}}CH_2(aq) = HO-CH_2-CH_2-OH(aq) \tag{15}$$

$$\qquad\qquad\quad \mathbf{2} \qquad\qquad\qquad\qquad \mathbf{1}$$

About 1.3 million tonnes of ethane-1,2-diol are produced each year for use as antifreeze and in the polyester fibre industry. The experimental rate equation for reaction 15 takes the form:

$$J = (k_0 + k_A[H^+] + k_B[OH^-])[C_2H_4O] \qquad (16)$$

At 30 °C, $k_0 = 9.2 \times 10^{-7}\,s^{-1}$

$$k_A = 1.7 \times 10^{-2}\,dm^3\,mol^{-1}\,s^{-1}$$

$$k_B = 1.7 \times 10^{-4}\,dm^3\,mol^{-1}\,s^{-1}$$

Which of the three terms in parentheses will dominate the right-hand side of the equation, and hence, what will the expression reduce to:

(i)   at high acidity (say pH = 1);

(ii) in neutral solution (pH = 7);

(iii) in basic solution (say pH = 14)?

From a consideration of these reduced forms, in which regions of pH do you think this reaction is catalysed and what are the catalysts?

In SAQ 2 you met two important examples of catalysis — **acid catalysis**, where the catalyst is H^+, and **base catalysis**, in which the catalyst is OH^-. Many reactions that occur in aqueous solution are susceptible to acid and/or base catalysis. Examples include many hydrolyses, decompositions, condensation reactions and molecular rearrangements. The experimental rate equation for such reactions may frequently be written as:

$$J = (k_0 + k_A[H^+] + k_B[OH^-])[X]^\alpha[Y]^\beta \qquad (17)$$

The three terms in parentheses in equation 17 reflect the fact that there are *three separate* mechanisms available for the overall conversion of reactants into products — an uncatalysed route, an acid-catalysed pathway involving H^+, and a base-catalysed route involving OH^-. The relative sizes of the rate constants (k_0, k_A and k_B), and the pH of the solution, will determine which mechanism predominates at the molecular level, and thus, which of the terms in equation 17 will dominate the rate equation.

2.3 Catalytic cycles

Before we go on to examine other catalytic systems, let's just pause to introduce a useful notation. The acid-catalysed hydrolysis of oxirane (C_2H_4O, **2**) is thought to proceed via the following mechanism:

$$\text{H}_2\text{C}-\text{CH}_2 + \text{H}^+ \underset{k_{-1}}{\overset{k_1}{\rightleftharpoons}} \text{H}_2\text{C}-\text{CH}_2 \qquad (18)$$

$$\text{H}_2\text{O} + \text{H}_2\text{C}-\text{CH}_2 \xrightarrow{k_2} \text{H}_2\overset{+}{\text{O}}-\text{CH}_2-\text{CH}_2-\text{OH} \qquad (19)$$

$$\text{H}_2\overset{+}{\text{O}}-\text{CH}_2-\text{CH}_2-\text{OH} \xrightarrow{k_3} \text{HO}-\text{CH}_2-\text{CH}_2-\text{OH} + \text{H}^+ \qquad (20)$$

These equations spell out the various steps along the pathway, but they do little to emphasize the catalytic nature of this reaction. A good pictorial way of achieving this is to use a **catalytic cycle**, such as that shown in Figure 1. There are a number

of points to note. The catalyst generally appears at the top of the cycle and the reactions are connected in a cyclic way such that one clockwise rotation from this point gives the sequence of reactions in the order in which they would appear in the 'conventional' representation of the mechanism, in this case the one shown in equations 18, 19 and 20. This means that during one journey round the circle the reactants (oxirane and water) are converted into product (ethane-1,2-diol), and the catalyst (H^+) is regenerated. The catalyst always remains *within* the cyclic pathway, whereas the reactants join the cycle at one point and leave, *as the product*, at another. Notice that the cycle does not show which steps are reversible. In this Block we shall adopt the convention of showing the reactants of catalytic cycles in bold type and the products in colour (as in Figure 1).

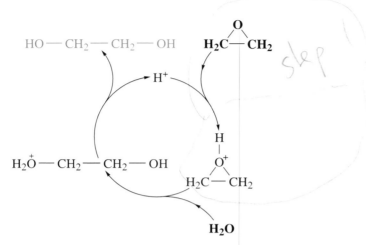

Figure 1   Catalytic cycle for the acid-catalysed hydrolysis of oxirane.

In this particular cycle, the forward reaction of step 1 in our mechanism (equation 18) is represented as follows:

$$(21)$$

In this step the reactant *enters* the cycle. The final step (equation 20) is written as

$$(22)$$

In this step the product *leaves* the cycle.

Figure 2 shows the catalytic cycle for the iodide-catalysed decomposition of hydrogen peroxide, which has the mechanism shown in equations 10 and 11:

$$H_2O_2 + I^- \xrightarrow{k_1} H_2O + IO^- \qquad (10)$$

$$IO^- + H_2O_2 \xrightarrow{k_2} H_2O + O_2 + I^- \qquad (11)$$

In *both* steps a reactant enters the cycle at the same time that a product leaves. This is represented (for the first step) as follows:

$$(23)$$

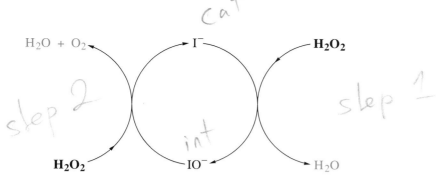

Figure 2   Catalytic cycle for the iodide-catalysed decomposition of hydrogen peroxide.

STUDY COMMENT   To gain some familiarity with catalytic cycles try the following SAQ.

SAQ 3   The non-complementary electron transfer (Block 3, Section 5.1) between Tl^I and Ce^{IV},

$$Tl^I + 2Ce^{IV} = Tl^{III} + 2Ce^{III} \tag{24}$$

is catalysed by Mn^{II} according to the catalytic cycle shown in Figure 3. Write down in the more 'conventional' way the three steps of this mechanism.

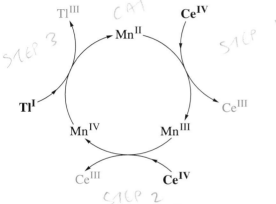

Figure 3   Catalytic cycle for the Mn^{II}-catalysed non-complementary electron transfer between Tl^I and Ce^{IV}.

2.4 Summary of Sections 1 and 2

1 It is convenient to divide catalysis into two types: homogeneous catalysis, in which reactants and catalyst are present together in the same phase, and heterogeneous catalysis, in which the catalyst is in a different phase from that of the reactants.

2 A catalyst is a substance that increases the rate of a chemical reaction without itself being consumed and without altering the position of thermodynamic equilibrium.

3 A catalyst increases the rates of the forward and back reactions in the same proportion.

4 A catalyst provides a reaction with an alternative pathway, involving a lower activation energy than the uncatalysed route. (The pre-exponential factor, A, may also change, but it does not usually have as much effect on the rate constant.)

5 A catalytic cycle shows a series of reactions connected in a cyclic fashion, such that during one journey round the cycle the reactants are converted into products and the catalyst is regenerated.

SAQ 4   Identify the most likely catalysts in the following catalysed reactions.

(a)   The reaction

$$NH_2NO_2(aq) = N_2O(aq) + H_2O(l)$$ (25)

which has the experimental rate equation: *acid* $(HNO_3?)$

$$J = k_R[NH_2NO_2][H^+]$$ (26)

(b)   The reaction

$$CH_3Br(aq) + H_2O(l) = CH_3OH(aq) + H^+(aq) + Br^-(aq)$$ (27)

which proceeds via the following catalysed mechanism:

$$CH_3Br + I^- \longrightarrow CH_3I + Br^-$$ (28) I^-

$$CH_3I + H_2O \longrightarrow CH_3OH + H^+ + I^-$$ (29)

(c)   The reaction

$$SO_2(g) + \tfrac{1}{2}O_2(g) = SO_3(g)$$ (30)

which has the catalytic cycle shown in Figure 4.

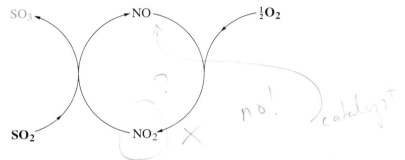

no!   catalyst

Figure 4   Catalytic cycle for the catalysed reaction between SO_2 and $\tfrac{1}{2}O_2$.

*this one confused me! I knew
that by convention the catalyst
appears at the top, but it
looked as if it was NO_2 that
was really driving the reaction.*

*However NO is regenerated
NO_2 isn't*

3 TRANSITION METAL CATALYSIS

We now turn our attention to one particularly important type of homogeneous catalysis, **transition metal catalysis**, in which the catalyst is a **transition metal complex**. These homogeneous systems often function effectively under milder conditions and are more selective than their heterogeneous counterparts. Table 2 includes a few examples of the industrial applications of homogeneous transition metal catalysis. This type of catalysis is also important in biological systems: transition metals such as iron, copper, molybdenum, manganese and cobalt are found in many enzymes and are thought to be intrinsic to their action.

But why should transition metals provide the essential ingredient in such a wide range of catalyst systems?

Table 2   Some industrial processes using homogeneous transition metal complex catalysts

Process*	Company	Reaction	Catalyst	Temperature/°C	Pressure/atm
alkene oligomerization	Shell	31	**3**/ethane-1,2-diol	100	80
hydroformylation	Union Carbide and others	32	$[RhH(CO)(PPh_3)_3]$ /organic solvent	60–120	15–25
carbonylation	(a) Monsanto	33	$[RhI_2(CO)_2]^-$/wet methanol	180	30–40
	(b) Tennessee–Eastman	34	$[RhI_2(CO)_2]^-$/ethanoic acid	180	30–40
hydrocyanation	Du Pont	35	$[Ni\{P(OC_6H_5)_3\}_4]$	70	a few atm

* Don't worry if you are unfamiliar with the terminology used in these processes. The reactions involved are listed below.

oligomerization $\qquad\qquad CH_2{=}CH_2 \longrightarrow C_{10}{-}C_{18}$ alkenes $\qquad$ (31)

hydroformylation $\qquad RCH{=}CH_2 + H_2 + CO \longrightarrow RCH_2CH_2CHO \qquad$ (32)

carbonylation $\qquad\qquad CH_3OH + CO \longrightarrow CH_3COOH \qquad$ (33)

$\qquad\qquad\qquad CH_3COOCH_3 + CO \longrightarrow CH_3COOCOCH_3 \qquad$ (34)

hydrocyanation $\quad CH_2{=}CH{-}CH{=}CH_2 + HCN \longrightarrow NC{-}CH_2{-}CH{=}CH{-}CH_3 \quad$ (35)

3.1 Why transition metals?

To understand why transition metals are such good catalysts, we must briefly survey some of the chemistry of these elements.

■ In terms of the electronic configurations of the transition metals, what distinguishes them from the main group elements?

▨ The transition elements usually have partially filled d shells.

The presence of partially filled d shells confers on the transition metals a range of interesting properties. To begin with, they form an amazing array of **coordination compounds**.

■ Coordination compounds were introduced in the Second Level Inorganic Course. The complex $[Ni(NH_3)_6]^{2+}$ has an *octahedral structure*. Try to draw out the structure of this complex. What are the coordinated ammonia molecules known as?

■ The structure of this complex is shown in Figure 5. The coordinated ammonia molecules are known as **ligands**.

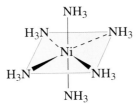

Figure 5 The structure of the $[Ni(NH_3)_6]^{2+}$ complex.

Transition metals are very 'social entities. They form coordination compounds with a whole range of different ligands, which can be either charged — for example, Cl^-, OH^-, CN^-, H^-, alkyl$^-$ (for example, CH_3^-) — or neutral, such as NH_3, H_2O, PR_3*, CO or alkene. It is this ability to form coordination compounds with almost any organic molecule that makes transition metals particularly useful industrial catalysts. For example, in 1964 the square-planar complex of rhodium(I), compound **4**, sometimes called *Wilkinson's catalyst*, was found to catalyse the hydrogenation of hex-1-ene to hexane:

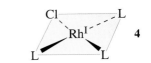

L = ligand, usually $P(C_6H_5)_3$

$$CH_3(CH_2)_3CH{=}CH_2 + H_2 = CH_3(CH_2)_4CH_3 \qquad (36)$$

The mechanism shown in Figure 6 (opposite) has been proposed to account for the kinetics of this reaction and the detection of certain intermediates. You will *not* be expected to reproduce such complex cycles; don't even worry too much about the chemistry of this process at the moment. The important things to notice are that first hydrogen, then the alkene become complexed to the rhodium; they react together while still attached to the transition metal, and then the product, the alkane, detaches itself from the complex and the catalyst is regenerated.

In coordination compounds such as **4–8**, the ligand and the central metal atom are *bonded* together. Thus, when a molecule or anion becomes coordinated to a metal, it will have *very* different properties from those of the corresponding free molecule or anion (the complexed metal also has different properties from the free metal ion).

■ The Second Level Inorganic Course introduced you to the relationship between the infrared stretching frequency of a bond and its force constant. The infrared stretching frequency of the $C{-}O$ bond in a free carbon monoxide molecule is $2\,143\ \text{cm}^{-1}$. When it is coordinated to iron, in the complex $[Fe(CO)_4]^{2-}$, the value drops to $1\,790\ \text{cm}^{-1}$. What conclusion can you draw about the strength of the $C{-}O$ bond in the coordinated and uncoordinated states?

■ The $C{-}O$ bond is weaker in the complex than it is in free carbon monoxide.

This weakening of the $C{-}O$ bond is a result of the formation of a bond between the *iron* atom and the *carbon* atom of the CO ligand. Similarly, when an alkene is coordinated to a transition metal, as in compound **7**, the bonding between the alkene and the rhodium atom weakens the carbon–carbon double bond. Coordination to the metal also alters the electron density in the alkene. The effect of weakening the double bond and altering the electron density often makes the alkene more susceptible to reaction, as in the conversion of compound **7** into compound **8**.

A second reason why transition metals make good catalysts is that they can adopt a range of stable **oxidation numbers** (or **oxidation states**).

---

*Just as ammonia and amines can be ligands, so can the corresponding phosphorus compounds. Phosphine, PH_3, is rarely used as a ligand since it inflames spontaneously in air. A much more common phosphorus ligand is triphenylphosphine, $P(C_6H_5)_3$ (or PPh_3), which is a lot easier to handle.

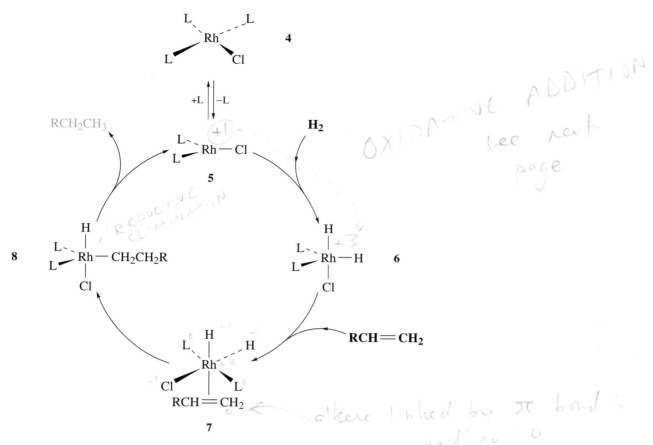

Figure 6 Possible catalytic cycle for alkene hydrogenation catalysed by $[RhClL_3]$.

- In SAQ 3 we said that Mn^{II} acts as a catalyst for the following non-complementary electron transfer:

$$Tl^{I} + 2Ce^{IV} = Tl^{III} + 2Ce^{III} \tag{37}$$

As you saw there, the catalysed reaction is thought to proceed via the following mechanism:

$$Ce^{IV} + Mn^{II} \longrightarrow Ce^{III} + Mn^{III} \tag{38}$$

$$Mn^{III} + Ce^{IV} \longrightarrow Ce^{III} + Mn^{IV} \tag{39}$$

$$Mn^{IV} + Tl^{I} \longrightarrow Mn^{II} + Tl^{III} \tag{40}$$

What property of manganese makes it a good catalyst for this reaction?

- Judging by the mechanism, manganese is a good catalyst for this reaction because it can adopt a range of stable oxidation states — Mn^{II}, Mn^{III} and Mn^{IV} in this case.

The uncatalysed mechanism of this reaction is similar to that of the non-complementary electron transfer between Tl^{III} and Fe^{II}, which we established in Block 3; by analogy

$$Tl^{I} + Ce^{IV} \longrightarrow Tl^{II} + Ce^{III} \tag{41}$$

$$Tl^{II} + Ce^{IV} \longrightarrow Tl^{III} + Ce^{III} \tag{42}$$

In this case, formation of Tl^{II} in the first step is rate limiting. By shuttling between the various oxidation states of manganese, the catalysed route avoids the formation of Tl^{II}, thus providing a lower-energy pathway between the reactants and the products.

Another example of this variation in oxidation number can be seen in the catalytic cycle given in Figure 6. First, let us remind ourselves of the rules for assigning oxidation states, as given in the Second Level Inorganic Course. For reference, these are reproduced in Box 1.

Box 1   Useful rules for the assignment of oxidation numbers

1   The oxidation number of a monatomic ion is equal to the charge on the ion. Thus, the oxidation numbers of iron and chlorine in $Fe^{2+}(aq)$ and $Cl^-(aq)$ are +2 and −1, respectively.

2   The oxidation number of an atom in its elemental form is zero. Thus, the values for aluminium and bromine atoms in $Al(s)$ and $Br_2(l)$, respectively, are zero.

3   The oxidation number of fluorine in compounds is always −1.

4   The oxidation number of oxygen in compounds is −2, except when it is bound to fluorine, or, as in compounds such as peroxides, bound to other oxygen atoms. Thus, in CO_2, the oxygen atom has an oxidation number of −2.

5   The oxidation number of hydrogen in compounds is taken to be +1, except in metallic hydrides such as NaH, where it is −1. Thus, in HCl, the oxidation number of hydrogen is +1.

6   The oxidation number of chlorine, bromine or iodine in compounds is −1, except in the compounds that they form with oxygen and with other halogen atoms. Thus, in HCl the oxidation number of chlorine is −1.

7   The sum of the oxidation numbers of the atoms in a compound or ion is equal to the charge on that compound or ion. Thus, HCl has no overall charge, and the oxidation numbers of H and Cl are +1 and −1, respectively.

■   What is the oxidation number of rhodium in $[RhCl_4]^{3-}$?

■   Since each of the chlorine ligands has an oxidation number of −1, and the total charge on the complex is −3, the rhodium must have an oxidation number of +1. Remember that the oxidation number is in no sense a measure of the actual charge carried by the rhodium atom.

As in this example, ligands are usually assigned an oxidation number equal to the charge on the 'free' molecule or ion; thus, OH^-, CN^- or alkyl ligands linked by a σ bond to a metal are all assigned an oxidation number of −1, whereas CO or $P(C_6H_5)_3$ have an oxidation number of 0. Alkenes (such as $RCH{=}CH_2$) have an oxidation number of 0 when they are linked to the transition metal through the π bond (as in structure **7** in Figure 6). However, such ligands can move within the complex and 'insert' themselves into another bond, as in the step that transforms **7** into **8** in Figure 6: here the alkene has inserted itself between rhodium and hydrogen, leading to the formation of an alkyl ligand. Formally, we have replaced a hydride ligand (oxidation number −1) by an alkyl ligand (oxidation number −1), so there is no change in the oxidation number of the rhodium in going from **7** to **8.** Links through π bonds followed by insertions are thought to be the mechanism by which many reactions of alkenes catalysed by transition metal complexes work.

Now concentrate on the following step in the catalytic cycle shown in Figure 6:

$$
\begin{array}{ccc}
& & H \\
& & | \\
L{\diagdown}Rh{-}Cl + H_2 & \longrightarrow & L{\diagdown}Rh{-}H \\
L{\diagup} & & L{\diagup}\ | \\
& & Cl
\end{array}
\qquad (43)
$$

5   **6**

In metal hydrides the oxidation number of hydrogen is taken to be −1; thus, in this step, the oxidation number of rhodium *increases* by two, from +1 to +3.

■   Is the rhodium oxidized or reduced in this process?

■   Since its oxidation number increases, the rhodium is oxidized.

The process shown in equation 43 is known as an **oxidative addition**. The rhodium(III) is converted back into rhodium(I) in the last step of the cycle —
that is, the change from compound **8** to compound **5**. This is known as a **reductive elimination**.

A third important aspect of transition metal chemistry should also be clear from equation 43: transition metals often show variable **coordination number**; in this case there is a change from three coordinate to five coordinate.

The final reason why transition metals make such good catalysts is that reactions in and around the complex are particularly susceptible to ligand effects. As you can see in Figure 6, as well as those ligands coordinated to the rhodium which participate in the mechanism (**participative ligands**), there are also a number of **non-participative ligands**. The latter influence the course of the reaction by modifying the structural and/or electronic properties of the complex. For example, if the chloride ligand in the complex **4** is replaced by either bromide or iodide the rate of hydrogenation of the alkene increases. This susceptibility of reactions to ligand effects provides the chemist with an invaluable means of tailoring the catalyst to suit individual processes.

> To summarize, transition metals make good catalysts for the following reasons:
>
> (a)   They form coordination compounds with a wide range of ligands.
>
> (b)   Coordination to a transition metal often makes a ligand more susceptible to reaction.
>
> (c)   They can adopt a range of oxidation numbers.
>
> (d)   They can adopt a range of coordination numbers.
>
> (e)   A catalytic reaction can be influenced by changing the ligands of the transition metal complex.

Having discussed *why* transition metal complexes make good homogeneous catalysts, we next turn our attention to some aspects of their industrial use.

3.2 Transition metal catalysis in action

The hydrogenation process shown in Figure 6 is not widely used in industry. A much more important application of transition metal catalysis is in the **oxo** or **hydroformylation*** process, which involves the addition of the units H and CHO to a carbon–carbon double bond to give an *aldehyde* (or *alkanal*):

$$RCH{=}CH_2 + CO + H_2 = RCH_2{-}CH_2{-}CHO \qquad (44)$$

To take a specific example, the annual production of butanal (**9**) by the hydroformylation of propene (equation 45) amounts to about 1.3 million tonnes:

$$\underset{\text{propene}}{CH_3CH{=}CH_2} + CO + H_2 = \underset{\textbf{9}}{CH_3CH_2CH_2CHO} \qquad (45)$$

Most of this butanal is used to make dioctylphthalate, which can be used to improve the physical properties of synthetic polymers.

---

* The name is derived from formaldehyde (methanal), HCHO.

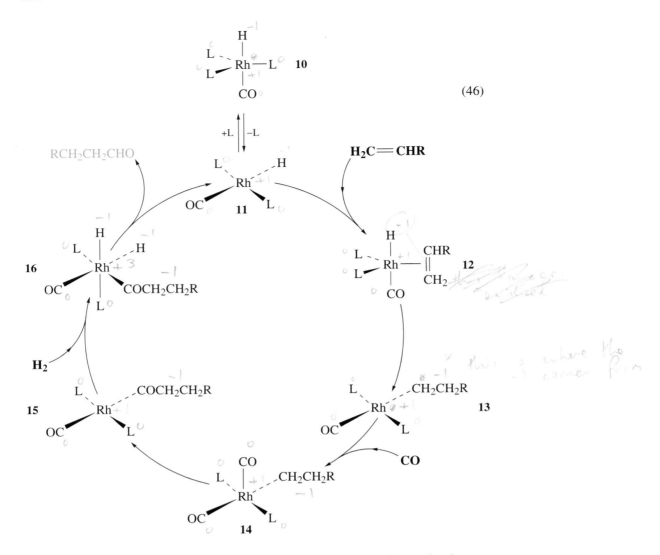

(46)

Figure 7   A possible catalytic cycle for the alkene hydroformylation reaction catalysed by [RhH(CO)L$_3$], where L is usually P(C$_6$H$_5$)$_3$.

Hydroformylation is catalysed by complexes of both cobalt and rhodium, but we shall concentrate here on the latter. A possible catalytic cycle for this reaction is shown in Figure 7. Don't worry too much about the stereochemistry of the various complexes in Figure 7. The important thing to concentrate on is how the reactants are converted into products.

SAQ 5   Write down the oxidation number of rhodium at each stage of the catalytic cycle in Figure 7, assuming that L = P(C$_6$H$_5$)$_3$. Is L a participative or non-partici-pative ligand?

To begin with, notice that it is not actually [RhH(CO)L$_3$] (**10**) that does the work; in solution, **10** partially dissociates to give **11**, which is the true catalyst. As before, the first step in the cycle involves the coordination of an alkene molecule to the rhodium through the π bond (**11** → **12**). In the next step (**12** → **13**), an internal rearrangement occurs; you can regard this either as the alkene inserting into the Rh—H bond or as Rh and H adding across the double bond of the alkene. This is followed by the addition of another CO molecule to the complex (**13** → **14**). The CO then inserts between the rhodium and the alkyl ligand (**14** → **15**).

■   What happens in the next stage (**15** → **16**)? How has the oxidation number of the rhodium changed?

■   The next stage involves an oxidative addition of hydrogen, to give the dihydride complex **16**. The oxidation number of the rhodium is increased by two, as you should have found in answering SAQ 5.

In the final stage ($16 \rightarrow 11$), the product aldehyde leaves when one of the hydrogens becomes bonded to the carbon of the C=O group. As a result, the rhodium(III) is converted back to rhodium(I); this is another example of a reductive elimination.

You may have noticed that there are two ways of adding the units H and CHO across the alkene double bond. It is the straight-chain form, **17**, that is the more useful, whereas the branched compound, **18**, is an unwanted side-product.

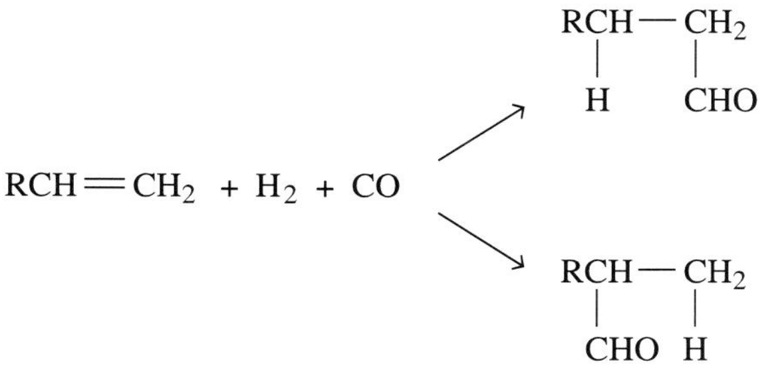

$$
\text{RCH}=\text{CH}_2 + \text{H}_2 + \text{CO} \quad
\begin{array}{l}
\nearrow \quad
\begin{array}{cc}
\text{RCH} - \text{CH}_2 & \textbf{17} \\
| \quad\quad | \\
\text{H} \quad\ \text{CHO}
\end{array} \\
\\
\searrow \quad
\begin{array}{cc}
\text{RCH} - \text{CH}_2 & \textbf{18} \\
| \quad\quad | \\
\text{CHO} \ \ \text{H}
\end{array}
\end{array}
\tag{46}
$$

If the catalyst is the complex **19** (**10** with L = CO), the selectivity of this process for the straight-chain product seldom exceeds 50%. However, if the triphenylphosphine complex **20** (**10** with L = P(C$_6$H$_5$)$_3$) is used, the selectivity increases to about 94%. This provides a clear example of how, by changing the ligands, a catalyst can be designed to carry out a particular reaction with the utmost efficiency. The increase in selectivity for the straight-chain compound on going from **19** to **20** can be explained in terms of a **steric effect**: the triphenylphosphine group is very bulky, and so the region around the rhodium atom in **20** is very crowded. As shown in Figure 8, of the two possible forms of the intermediate **13** (with L = P(C$_6$H$_5$)$_3$), it is clear that the ligands in the straight-chain form (**21**) will take up less room around the rhodium than those in the branched-chain form (**22**), and thus will be less crowded. This causes **21** to be formed in preference to **22**.

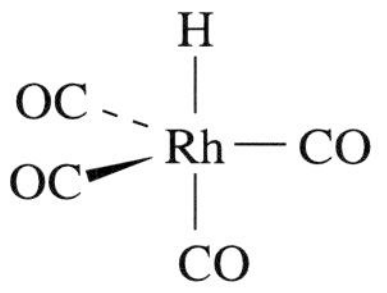

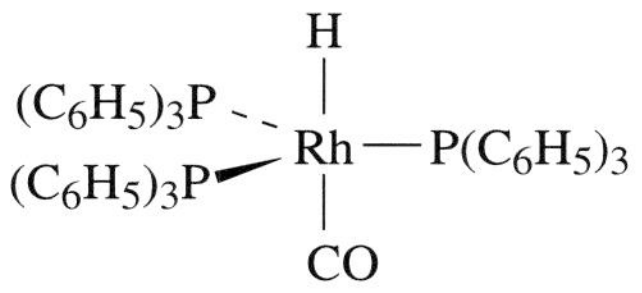

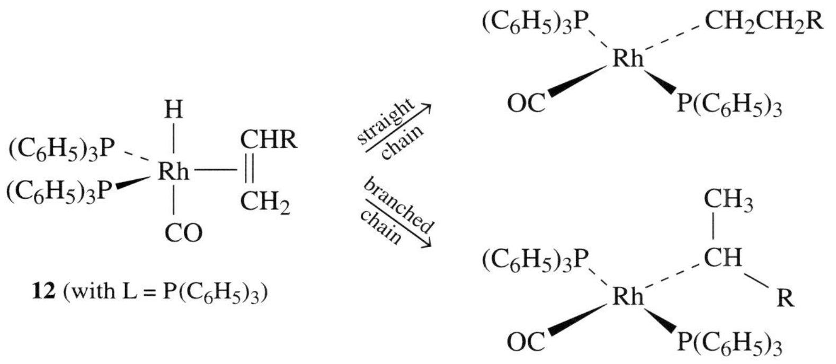

Figure 8 Possible outcomes of the internal rearrangement (**12** → **13**) during the hydroformylation reaction shown in Figure 7, with L = P(C$_6$H$_5$)$_3$.

The only other catalytic system that might be employed in this type of process is a cobalt complex such as **23**, but this is by no means as efficient. The rhodium catalyst speeds the reaction up by a factor of between 100 and 1 000 times more than the cobalt system. Moreover, it operates at lower pressures and temperatures (thus saving energy), and it has a higher selectivity for the straight-chain aldehyde product. However, rhodium is about 3 500 times more expensive than cobalt! This means that catalyst loss *must* be kept to a minimum. The amount of catalyst employed in such systems is of the order of 0.1 to 0.01 g for every kilogram of alkene converted; if, in this conversion, there were a loss of only about 3×10^{-5} g, this would correspond to an increase in the price of the product by a factor of 10–15%.

[CoH(CO)$_2$L] **23**

L = CO or P(C$_4$H$_9$)$_3$

There is another important difference between the rhodium and the cobalt systems. The rhodium catalyst gives about a 96% yield of the aldehyde, but the cobalt complex can catalyse a further reaction with hydrogen to give an alcohol (alkanol), leading to a reduced yield of aldehyde:

$$
\text{RCH}=\text{CH}_2 + \text{CO} + \text{H}_2 \longrightarrow \text{RCH}_2\text{CH}_2\text{CHO} \xrightarrow{\text{H}_2} \text{RCH}_2\text{CH}_2\text{CH}_2\text{OH}
\tag{47}
$$

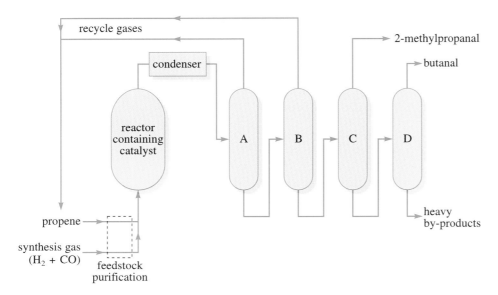

Figure 9 Simplified flow diagram for the hydroformylation process catalysed by rhodium complexes with tertiary phosphine ligands.

Under the conditions of the reaction the rhodium catalyst will not catalyse the hydrogenation in the second step to any great extent. Hence, the rhodium catalyst will be preferred if the aldehyde is the desired product. However, if the alcohol is required, a one-step cobalt process may be more desirable than a multi-step process, which involves making the aldehyde, separating and purifying it, and then hydrogenating it in a separate step. As you will see again in Block 5, in the final analysis, the economics of the two processes is the deciding factor.

Figure 9 shows the flow diagram for the rhodium-catalysed process that is operated by Union Carbide at Ponce, Puerto Rico. As with many other processes that you have met, one of the starting materials is synthesis gas. However, before it can be used in this process it has to be purified. This involves removing hydrogen sulfide and carbonyl sulfide, which are good ligands for transition metals. These sulfides complex to the rhodium in preference to the hydrogen and the alkene, thereby reducing the concentration of active catalyst. A process such as this, in which the catalyst is rendered inactive, is known as **poisoning**. We shall have more to say about this in Block 5.

The stream of purified H_2 and CO is mixed with propene, and the mixture enters the reactor at its base. The reactor contains the rhodium catalyst dissolved in a liquid phase, which is kept at 80–120 °C under a pressure of 15–25 atm. The reactants dissolve in the liquid, where the reaction takes place. The product evaporates from the top of the reactor. This effluent is then condensed into a liquid–vapour separator (A) to give a liquid that contains appreciable amounts of propene. A product-stripping distillation column (B) is used to separate this and recycle it. The remaining liquid then goes on to two further distillation columns (C and D) to remove the branched-chain product, 2-methylpropanal (**18**; R = CH_3) from the straight-chain product, butanal (**17**; R = CH_3).

A significant problem in many industrial applications of homogeneous catalysis, although not such a difficulty in the process described above, is that of separating the products from the (often valuable) catalyst. In heterogeneous catalysis the solid catalyst is readily separated from gases or from liquids by decantation or filtration. In homogeneous catalysis it is frequently necessary to distil the reaction mixture in a separate step so that the catalyst can be recycled. This is often very expensive in energy. Attempts are being made to combine some of the advantages of homogeneous and heterogeneous catalysis to overcome such problems of separation: here we look briefly at two general strategies.

1 Supported catalysts

Considerable research effort has gone into 'supporting' transition metal complex catalysts by attaching them to polymers. The polymer, often a polystyrene cross-linked with 1–2% *para*-divinylbenzene (to make it less soluble in organic solvents), is treated to attach a diphenylphosphine unit, and this acts as a ligand for the transition metal complex, as, for example, in structure **24**. Such polymers will not dissolve in organic solvents (though some will swell). However, the transition metal complex on the surface of the polymer can protrude into the solution so that it is readily available for catalysis. Whether you regard these as homogeneous or heterogeneous systems is up to you! A modification of this approach is to support the transition metal complex on silica or alumina by treating the surface hydroxyl groups to produce a ligand capable of coordinating with a transition metal. Removal of the catalyst after the reaction is then a simple matter, since the catalyst is a solid.

$$H_2C{=}CH{-}\hexagon{-}CH{=}CH_2$$

para-divinylbenzene

polymer — $\hexagon$ — P — Rh — Cl   **24**

with P bearing C_6H_5, C_6H_5 substituents and Rh bearing $P(C_6H_5)_3$, $P(C_6H_5)_3$ ligands.

2 Water-soluble catalysts

An alternative strategy has been the development of water-soluble transition metal catalysts. One of these is used commercially in the Ruhrchemie–Rhone Poulenc hydroformylation process. The catalytic cycle is very similar to that in Figure 7, but the ligand attached to the rhodium has been modified (**25**) by addition of a sulfonate group, and this makes the catalyst soluble in water. The other components are dissolved in an organic liquid that is immiscible with water (such as toluene), and the reaction occurs at the water/toluene interface; the product aldehyde is virtually insoluble in water. Recovery of the catalyst is straightforward: the two liquid phases are easily separated and there is far less loss of rhodium than in the traditional single-phase hydroformylation.

$$P{-}{\left[\hexagon{-}SO_3^- \ Na^+\right]}_3 \qquad \textbf{25}$$

3.3 Polymerization

In terms of volume, the polymerization of ethene, propene and other vinyl monomers using transition metal complex catalysts is one of the most extensive uses of these catalysts. The first systems, developed in the 1950s, were the *Ziegler–Natta catalysts* and the *Phillips catalysts*. The Ziegler–Natta catalysts are based on mixed titanium/aluminium complexes, which appear to be soluble in hydrocarbon solvents (though there is some doubt whether they are truly soluble). The Phillips catalysts are based on chromium compounds attached to silica, and are definitely heterogeneous systems. New generations of Ziegler–Natta catalysts use titanium compounds supported on $MgCl_2$, or incorporate vanadium compounds. These result in processes that can be run at lower temperatures; indeed, the catalyst's activity is so high that the catalyst residue can be left in the polymer, saving an expensive separation stage. The vanadium-based catalysts do seem to be soluble, and are particularly useful in the production of random *copolymers* such as the ethene–propene (ethylene–propylene) copolymers, which are often used as synthetic rubbers. (Copolymers are polymers made from more than one monomer. The monomer units can be arranged in various ways: alternating, in blocks, or randomly along the chain.)

Although the mechanisms of these reactions have received a lot of attention, many are still not fully understood. It is thought that they usually include a step in which the incoming alkene links to the metal through its π bond, and then inserts into the bond between the metal and the growing polymer chain.

3.4 Summary of Section 3

1   Transition metals provide an essential ingredient of many of the catalysts used both in industry and in nature.

2   Transition metals make good catalysts because:

(a) they form complexes with a wide range of different ligands, including many organic molecules; complexation of a reactant such as an alkene to a metal centre often makes the compound more susceptible to reaction;

(b) they can adopt a range of stable oxidation numbers;

(c) they can adopt a range of coordination numbers;

(d) reactions of a transition metal complex can be influenced by the nature of the ligand, and this allows the catalyst to be tailored to achieve high selectivity for the desired transformation.

3   Hydroformylation involves the addition of the units H and CHO to a double bond. This process can be catalysed by rhodium and cobalt complexes. The use of bulky triphenylphosphine ligands increases the selectivity for the desired straight-chain product.

4   Separation of the catalyst from the reaction mixture is often a problem with homogeneous catalysis. Attaching the complexes to a solid support enhances the ease of separation. Another recent development has been the use of two-phase liquid systems, in which the catalyst is dissolved in water and the other components are dissolved in an organic solvent: here, catalyst recovery is straightforward.

5   One of the most important industrial applications of transition metal complex catalysts is in the polymerization of alkenes. Homogeneous systems include new generations of highly active Ziegler–Natta-type catalysts, based on mixed titanium/vanadium complexes.

SAQ 6   One of the stages in the manufacture of Nylon-6,6 involves the production of pent-3-enenitrile (**27**) by the hydrocyanation (Table 2) of 1,3-butadiene (**26**) in the presence of a nickel complex catalyst.

$$CH_2{=}CH{-}CH{=}CH_2 + HCN = CH_3{-}CH{=}CH{-}CH_2{-}CN \qquad (48)$$

$$\mathbf{26} \qquad\qquad\qquad\qquad\qquad \mathbf{27}$$

A possible mechanism is shown in the catalytic cycle in Figure 10.

(a) Identify the participative and non-participative ligands.

(b) In which steps do the reactants enter the cycle, and in which does the product leave?

(c) Which step involves an oxidative addition? What is the change in the oxidation number of the nickel during this step?

(d) In which step does the coordination number of nickel change from five to four?

SAQ 7   Which of the following statements is/are **true**?

(a) Homogeneous transition metal catalysis is particularly susceptible to changes in the nature of the ligands.

(b) Transition metal ions form complexes with few ligands.

(c) Hydroformylation involves the addition of the units H and CHO across a double bond.

(d) Homogeneous catalysis is preferred over its heterogeneous counterpart because it is easier to remove the catalyst in the former process.

(e) Poisoning is the process whereby the catalyst is rendered inactive.

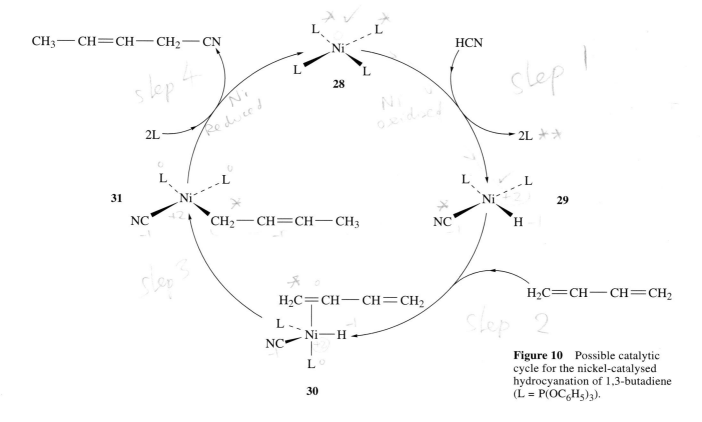

Figure 10 Possible catalytic cycle for the nickel-catalysed hydrocyanation of 1,3-butadiene (L = P(OC$_6$H$_5$)$_3$).

4 THE ENZYMES

In the last few Sections you have come across a number of synthetic catalysts; we now turn our attention to natural catalysts — the **enzymes**. These substances mediate (catalyse) the chemical reactions of all living organisms; they break down food, build up body material, maintain life processes and control energy storage and conversion. Yet, they do all this chemistry at about 37 °C, at about pH 7 in aqueous solution, and under a pressure of 1 atm. Clearly, we have much to learn from the enzymes!

■ Look back at Table 1 in Section 2.2. What is the most striking feature of the enzyme-catalysed decomposition of hydrogen peroxide in comparison with the other catalysed processes?

▨ The reaction catalysed by catalase goes *so much* faster than the decompositions catalysed by I$^-$ and Fe^{2+} — a factor of up to 10^6 faster. A glance at the activation energies shows how this comes about. For the enzyme-catalysed process the activation energy is reduced to only 7 kJ mol^{-1}.

In general, non-enzyme catalysts rarely come within a factor of a million of matching enzyme activity. From our point of view, a major question is whether this extraordinary behaviour can be analysed using the principles we have developed so far.

But, before we get involved in the kinetics and mechanisms of enzyme action, let us dwell briefly on the structure of enzymes, and the nature of enzyme-catalysed reactions. The hydrogenation of pyruvate (**32**) to lactate (**33**) is a typical enzyme-catalysed reaction (*note* NADH is a *reactant*, not the enzyme):

$$\text{NADH} + \text{H}^+ + \text{H}_3\text{C}-\underset{\underset{\textbf{32}}{\overset{\|}{\text{O}}}}{\text{C}}-\text{COO}^- = \text{NAD}^+ + \text{H}_3\text{C}-\underset{\underset{\textbf{33}}{\overset{|}{\text{OH}}}}{\text{CH}}-\text{COO}^- \qquad (49)$$

In the rest of this Block you will meet a number of organic compounds and complex biological molecules, such as NADH and NAD^+. However, you do not have to worry too much about their structures and names; *you will not be expected to remember them*. The important thing to take note of is the *change* that occurs in the reaction. In equation 49, for example, two hydrogen atoms are added across a carbon–oxygen double bond to form an OH group.

As you can see, this is *not* a simple hydrogenation — involving a molecule of hydrogen as such; in this case, the two hydrogens are provided by a hydrogen ion (H^+) and by the complex biological molecule nicotinamide adenine dinucleotide, which is represented in its reduced form by the abbreviation NADH (Figure 11).

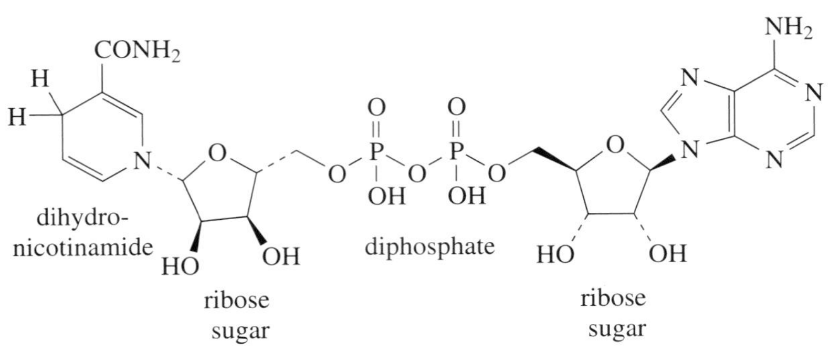

Figure 11 The structure of nicotinamide adenine dinucleotide (NADH).

We can represent the involvement of NADH as follows:

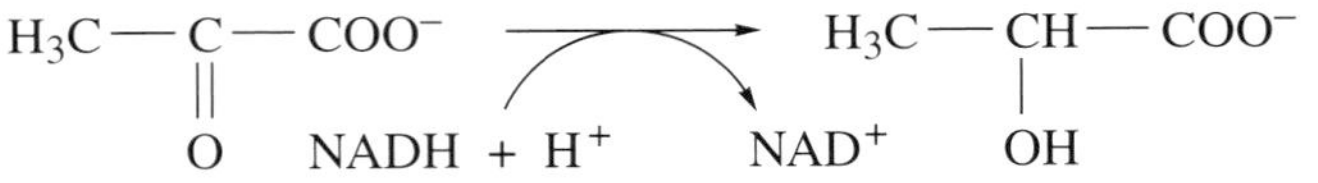
(50)

Unlike the arrows you met in the catalytic cycles in Sections 2 and 3, the horizontal arrow in equation 50 is *not* an indication that this is an elementary process. It is used to symbolize the overall conversion of one compound into another. As you may know from other courses, this is a widespread usage in biochemistry, especially in the representation of metabolic pathways such as that in Figure 12. This shows part of the glycolytic pathway, the series of chemical reactions by which many organisms oxidize glucose, an 'organic fuel'. Each stage in this metabolic pathway is catalysed by a different enzyme and each involves a number of elementary steps.

glucose ⟶ glucose 6-phosphate ⟶ fructose 6-phosphate

↓ six steps

lactate ⟵ pyruvate ⟵ phosphoenol pyruvate

33 **32**

Figure 12 Part of the glycolytic pathway.

The conversion of pyruvate into lactate is the last stage of the glycolytic pathway, and is catalysed by the enzyme *lactate dehydrogenase*. This is an extremely large molecule: it has a molar mass of about 36 000 g mol^{-1} and contains over 5 000 atoms. Enzymes are essentially proteins; that is, they are naturally occurring macromolecules consisting of a large number of units, or 'residues', derived from **amino acids** (**34**). These amino acids are covalently linked together to form long unbranched chains, as shown in Figure 13. There are twenty common amino acids (their structures are listed in the S342 *Data Book*), each with a different side-chain, R. All the amino acids (except glycine, in which the group R in **34** is a hydrogen atom) contain a carbon atom attached to four different groups. They are thus chiral; that is, they have a non-superimposable mirror image. However, only one optical isomer occurs naturally. The bonds between the amino acid residues are called peptide bonds, and this is why smaller proteins are often called polypeptides.

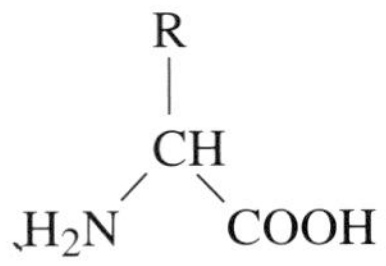

34, an amino acid

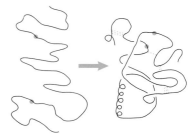

Figure 13 A section of protein chain. (R^1, R^2, etc., represent the side-chains of different amino acid residues.)

It is the order of the residues in the chain that differentiates one protein from another: this sequence is known as the *primary* structure. One of the essential features of an enzyme is that it has a precise three-dimensional shape. This is dictated by the primary structure, which causes the polypeptide chain to fold up in a highly exact way, which is characteristic of the particular enzyme; it is shown schematically in Figure 14. This folding pattern is known as *higher-order* structure, and it means that every single molecule of the enzyme will have precisely the same convoluted three-dimensional structure. The enzyme is held in this shape by hydrogen-bonding, and a number of other intramolecular interactions. Some of these are indicated schematically by the coloured lines in Figure 14. As well as the protein part of the enzyme, such species also contain other molecules, called *prosthetic groups* or *coenzymes*, which often involve metal ions.

A 'picture' of the enzyme lactate dehydrogenase, which catalyses the conversion of pyruvate into lactate (equation 49), is shown in Figure 15. The reactant sits inside a pocket in the enzyme, which is known as the **active site**. By arranging the right amino acid side-chains in the correct position around the pyruvate, the enzyme supplies the perfect environment for the reaction, thus providing a route from reactants to products that proceeds with a very low activation energy. If the enzyme is to furnish the right environment for reaction, the active site must be specifically suited for the reactant. Thus, although lactate dehydrogenase catalyses the reduction of pyruvate (**32**), it fails to catalyse the reduction of the apparently very similar substance amino oxoacetate (**35**). This is because the active site of the enzyme is not correct for the latter molecule.

Figure 14 The formation of an enzyme by the folding of a protein chain. Notice how two points widely separated in the chain can be brought close together by folding.

$$H_3C - \underset{\underset{O}{\|}}{C} - COO^-$$

pyruvate, **32**

$$H_2N - \underset{\underset{O}{\|}}{C} - COO^-$$

35

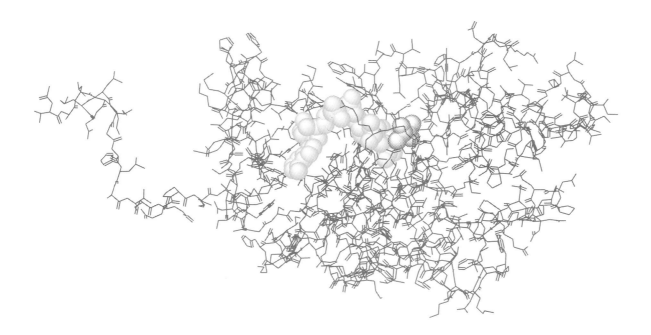

Figure 15 A computer-generated depiction of the enzyme lactate dehydrogenase. The blue area represents pyruvate (the substrate) and the grey area represents the coenzyme.

Because the active site is generally appropriate for only one reaction, enzyme-catalysed reactions have a high selectivity. For example, lactate dehydrogenase catalyses the formation of only *one* of the optically active forms of lactate, **36** and **37**:

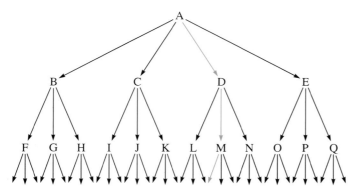

There are a great many enzymes in the human body, each catalysing a different specific reaction. Their catalytic efficiency, coupled with their selectivity, has extremely important consequences. As we saw with synthesis gas, an individual reactant may undergo a great number of reactions, all of which are thermodynamically favourable. In turn, the products of these reactions can react in many different ways, and so can *their* products (Figure 16). Enzymes, by virtue of their selectivity, will catalyse only one process, such that a series of enzymes will enhance *one particular* pathway, the **metabolic pathway**, through the maze of possible reactions. The body can switch these pathways on or off as necessary, by controlling the concentration and activity of the various enzymes involved.

Figure 16 The ability of a series of enzymes to select one of a large number of reaction pathways. Each arrow represents a possible chemical reaction. The catalysed pathway is shown in colour.

4.1 Enzyme kinetics

In this Section we shall examine the kinetics and mechanism of enzyme action. The reaction we shall study is the conversion of malate (**38**) into fumarate (**39**), a reaction that involves the elimination of water to form a carbon–carbon double bond:

This reaction forms part of the 'tricarboxylic acid cycle', which in turn is part of the process whereby glucose is ultimately oxidized to carbon dioxide and water. The details need not concern us; suffice to say that the process in equation 52 is catalysed by the enzyme *fumarase*, which has a molar mass of about $200\,000\,\text{g mol}^{-1}$. For a number of reasons, it is not easy to study enzyme reactions in living systems (*in vivo* studies). For example, the reactants and products under study are usually also reactants for other enzymes, and this will complicate the analysis. We are thus forced to study such reactions under artificial conditions (*in vitro* studies). Fumarase, which can be obtained in a crystalline form from pig heart, has been found to catalyse this reaction *in vitro*. However, it must be remembered that this is at best a poor approximation to the *in vivo* environment, and comparisons between the two must be made with care.

In enzyme-catalysed reactions the reactant is known as the **substrate**. Because this reaction involves only one reactant, fumarase is known as a *one-substrate enzyme*.

■ Following the procedures outlined in Block 2, the first stage in the analysis of this enzyme-catalysed reaction is to make a guess at the experimental rate equation. What is its most likely form? (This reaction exhibits time-independent stoichiometry.)

■ From the stoichiometry of the reaction and the fact that it is catalysed, the rate could be dependent on the concentration of the malate (because this is the substrate, it will be referred to by the symbol S) and that of the enzyme, E. So, a reasonable assumption would be:

$$J = k_R[S]^\alpha[E]^\beta \tag{53}$$

In principle, it should be possible to measure the appropriate kinetic reaction profiles and use the differential or integration methods to determine α and β. However, this is rarely possible, and this highlights one of the problems of enzyme kinetics. Not only are enzymes very complex, but they are also very delicate, and the molecules can easily become somewhat distorted in shape, and thus catalytically inactive. This process, known as **denaturation**, occurs especially easily if the enzyme is used in an environment other than that for which it has evolved. Because kinetic experiments are carried out *in vitro* rather than *in vivo*, an enzyme can become denatured, even *during* a reaction. For this reason, and others that we shall not go into here, enzyme kinetic studies usually involve measurements of the initial rates, so that there is insufficient time for these effects to become important.

■ Table 3 lists data for the initial rate of reaction 52 as a function of the concentration of the enzyme*, but with the *same* concentration of the substrate. How could these data be used to determine the value of β?

■ If the differential method is used, a plot of log J_0 versus log $[E]_0$ should be a straight line with a slope equal to the partial order with respect to the enzyme, β.

A plot of log J_0 versus log $[E]_0$, based on the data in Table 3 is shown in Figure 17. In this case, a straight line is indeed obtained and the slope is 0.99. We can therefore conclude that this reaction is most likely first order in the enzyme; that is, $\beta = 1$.

Figure 17 The variation in the log of the initial rate with the log of the enzyme concentration for the conversion of malate into fumarate at 293 K and pH 7 ([malate] = 0.01 mol dm^{-3} throughout).

Table 3 Kinetic measurements for the enzyme-catalysed conversion of malate into fumarate at 293 K and pH 7; [S] = 0.01 mol dm^{-3} throughout

$[E]_0$	J_0
10^{-10} mol dm^{-3}	10^{-7} mol dm^{-3} s^{-1}
5	1.1
10	2.1
15	3.2
20	4.2
25	5.3

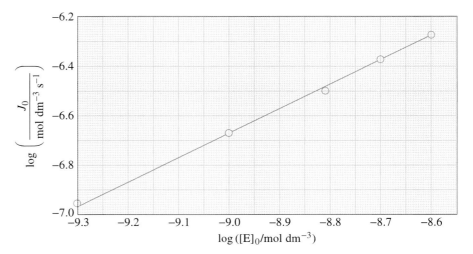

* It is not always possible to calculate the concentration of an enzyme from the amount of it in solution and its molar mass. This is because they are not always isolated 100% pure. Biochemists have a number of ways of getting around this, but such methods need not concern us in this Course.

Table 4 Kinetic measurements for the enzyme-catalysed conversion of malate into fumarate at 293 K and pH 7; $[E]_0 = 1.6 \times 10^{-9}$ mol dm^{-3} throughout

$\dfrac{[S]_0}{10^{-3} \text{ mol dm}^{-3}}$	$\dfrac{J_0}{10^{-8} \text{ mol dm}^{-3} \text{ s}^{-1}}$
0.1	2.5
0.5	9.5
1.0	15.3
2.0	22.0
4.0	28.3
7.5	33.0
10.0	34.5

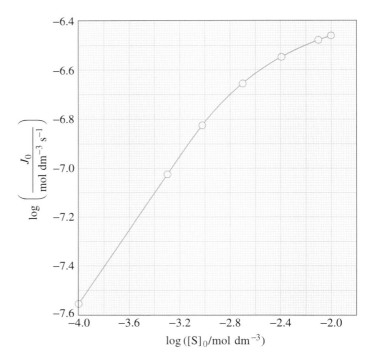

Figure 18 The variation in the log of the initial rate with the log of the concentration of the substrate, S, for the conversion of malate into fumarate at 293 K and pH 7 ($[E]_0 = 1.6 \times 10^{-9}$ mol dm^{-3} throughout).

Table 4 lists data for the initial rate of this reaction as a function of the concentration of the substrate, *with the same concentration of enzyme*. The partial order with respect to the substrate, α, should be obtainable from the corresponding plot of $\log J_0$ versus $\log [S]_0$: the plot is shown in Figure 18.

■ What is the problem with this plot?

▨ The plot in Figure 18 is *not* a straight line. The partial order with respect to the substrate, S, which is given by the slope of the graph, seems to decrease as the concentration of the substrate is increased. In fact, a more detailed experimental analysis reveals that the partial order changes from 1 to 0.

The chemical rate equation that is consistent with the experimental data is thus

$$J_0 = k_R [E]_0 [S]_0^\alpha \qquad (54)$$

where α varies from 1 to 0, depending on the concentration of the substrate. This type of chemical rate equation is characteristic of enzyme-catalysed reactions, and was first recognized during the latter part of the nineteenth century.

Chemists soon realized that the strange behaviour of the initial rate, with changing substrate concentration, can often be expressed mathematically as follows:

$$J_0 = \frac{k_R [E]_0 [S]_0}{K_m + [S]_0} \qquad (55)$$

where K_m is a constant. Since the mode of studying enzyme action is to measure the initial rates at a *constant* enzyme concentration while the substrate concentration is varied, the expression can be reduced to

$$J_0 = \frac{V[S]_0}{K_m + [S]_0} \qquad (56)$$

where

$$V = k_R[E]_0 \qquad (57)$$

Equation 56 is known as the **Michaelis–Menten* equation**, and the constant K_m is called the **Michaelis constant**. We shall discuss the significance of V and K_m shortly.

---

* It was the French chemist Victor Henri who in 1908 first proposed this type of equation and the mechanistic interpretation that we shall discuss in Section 4.4. However, this was refined in 1913 by the German American chemist Leonor Michaelis and his Canadian assistant Mary L. Menten.

First, let us examine how well the data in Table 4 are accommodated by this chemical rate equation. Because the concentration of the enzyme is constant, we shall employ the reduced form, equation 56.

■ When we applied the differential method to the data in Table 4, we found that at low substrate concentrations the reaction is first order in substrate, but at high substrate concentrations the reaction becomes zeroth order in the substrate. Is equation 56 consistent with this finding?

■ Yes. K_m will dominate the denominator when $[S]_0$ is much smaller than K_m, and equation 56 then reduces to

$$J_0 = \frac{V[S]_0}{K_m} \tag{58}$$

which is first order in the substrate. When $[S]_0$ is much greater than K_m, $[S]_0$ will dominate the denominator, and equation 56 reduces to

$$J_0 = V \tag{59}$$

which is zeroth order in the substrate.

This argument is similar to the logic applied in Block 3 for comparing a theoretical and an experimental rate equation. Thus, equation 56 seems to be *qualitatively* reasonable, but, before we can be absolutely certain that this is the correct chemical rate equation, we must fit the data in Table 4 to equation 56 in a *quantitative* fashion. We shall use the data to calculate values of K_m and V, and then employ a graphical method to see how well the curve generated by equation 56 accords with the data.

■ How can we determine the value of V?

■ Equation 59 indicates that at high substrate concentrations V is equal to J_0.

In fact, V is defined as the **limiting** (or maximum) **rate** at high substrate concentrations.

■ Figure 19 shows a plot of J_0 against $[S]_0$ over an extended range ($[E]_0$ is again kept constant). What is the value of V?

■ At high substrate concentrations, the initial rate takes a limiting value of about $39 \times 10^{-8}\ \text{mol dm}^{-3}\ \text{s}^{-1}$, so this is our value for V.

As the substrate concentration is raised, the initial rate reaches a limiting value V, as shown in Figure 19.

■ What does the limiting rate signify in terms of the enzyme?

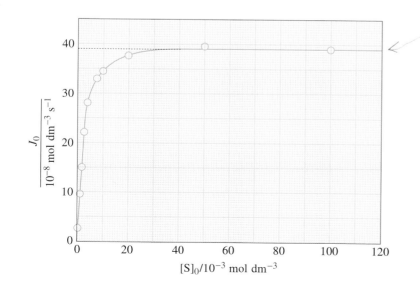

38 × 10⁻⁸ mol dm⁻³ s⁻¹

Figure 19 A plot of the initial rate, J_0, versus $[S]_0$ for the enzyme-catalysed reaction 52 ($[E]_0 = 1.6 \times 10^{-9}\ \text{mol dm}^{-3}$ throughout). *Note This is not a reaction profile.*

▩   Under these circumstances, all the catalytic sites are filled (the enzyme is saturated), and as fast as the sites become free through the departure of the reaction
product(s), they are refilled by incoming substrate.

▩   What happens to equation 56 when the initial rate of reaction, J_0, is half the
limiting value V? What do you conclude about the significance of the constant
K_m? [*Hint*   Substitute $J_0 = V/2$ into equation 56, and then rearrange this to give
an expression for K_m.]

▩   If $J_0 = V/2$, then equation 56 becomes

$$\frac{V}{2} = \frac{V[S]_0}{K_m + [S]_0} \qquad (60)$$

which rearranges to

$$V(K_m + [S]_0) = 2V[S]_0 \qquad (61)$$

and thence

$$K_m + [S]_0 = 2[S]_0 \qquad (62)$$

or

$$K_m = [S]_0 \qquad (63)$$

In other words, K_m *is equal to the concentration of the substrate that gives half
the limiting rate.* This is the usual operational definition of K_m.

STUDY COMMENT   You should now attempt the following SAQ, and check the answer before
reading on.

SAQ 8   Using the data in Table 4, plot a graph of J_0 against $[S]_0$. Determine the
value of K_m by finding the value of $[S]_0$ when J_0 has a value of $V/2$. (Assume $V =
39 \times 10^{-8}\,\mathrm{mol\,dm^{-3}\,s^{-1}}$.)

Now that we have values for K_m ($1.6 \times 10^{-3}\,\mathrm{mol\,dm^{-3}}$) and V ($39 \times 10^{-8}\,\mathrm{mol\,dm^{-3}\,s^{-1}}$),
we can compare the curve generated by equation 56, based on these values, with the
experimental data. In fact, this is how we drew the curve through the points in Figure
27 (in the answer to SAQ 8), which is reproduced here as Figure 20. Clearly, there is
close correspondence between the points and the line, so equation 56 is acceptable on
both qualitative and quantitative grounds.

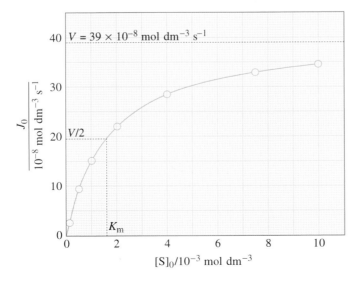

Figure 20   A plot of J_0 versus $[S]_0$ for the enzyme-catalysed reaction 52, using the data in
Table 4 ($[E]_0 = 1.6 \times 10^{-9}\,\mathrm{mol\,dm^{-3}}$ throughout).

Although this seems a reasonable method of obtaining the values of K_m and V, a much better way of evaluating these parameters would be via a *linear* graphical method. You met a similar argument in Block 2. The rate constant k_R is usually obtained from a linear plot of a function of the concentration against time, rather than using the half-life method.

Equation 56 can be converted into the form of an equation for a straight line as follows. First, it is inverted, to give equation 64:

$$\frac{1}{J_0} = \frac{K_m + [S]_0}{V[S]_0} \tag{64}$$

which can be written as

$$\frac{1}{J_0} = \frac{K_m}{V[S]_0} + \frac{[S]_0}{V[S]_0} \tag{65}$$

which reduces to the following:

$$\frac{1}{J_0} = \frac{K_m}{V}\left(\frac{1}{[S]_0}\right) + \frac{1}{V} \tag{66}$$

Equation 66 tells us that a plot of $1/J_0$ against $1/[S]_0$ should be a straight line with a slope equal to K_m/V and an intercept (when $1/[S]_0 = 0$) equal to $1/V$. This is known as a **Lineweaver–Burk plot**.

STUDY COMMENT   To use equation 66, you must be confident about interpreting the labels on the axes of a graph involving a plot of one reciprocal quantity ($1/J_0$) against another ($1/[S]_0$). The following SAQ gives you a chance to check this. Make sure you try it at some stage.

SAQ 9   Figure 21 shows a Lineweaver–Burk plot from the data in Table 4. Use this Figure to determine the values of V and K_m.

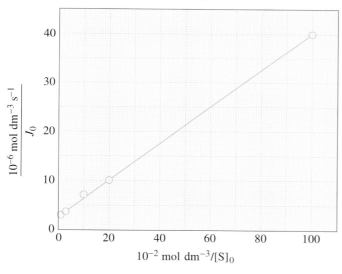

Figure 21   A Lineweaver–Burk plot of $1/J_0$ against $1/[S]_0$ using the data in Table 4.

Notice one feature of the plot in Figure 21: most of the data are clustered at low values of $1/[S]_0$. This gives undue weight to the measurements made at low substrate concentrations, which are often the least accurate. To overcome this problem, equation 56 can be 'organized' in different ways to give two other linear forms. These are the **Hanes plot**:

$$\frac{[S]_0}{J_0} = \left(\frac{1}{V}\right)[S]_0 + \frac{K_m}{V} \tag{67}$$

and the **Eadie plot**:

$$J_0 = -K_m\left(\frac{J_0}{[S]_0}\right) + V \tag{68}$$

Each type of plot has its own particular advantages and disadvantages. From a statistical point of view, the Hanes plot is the most 'well balanced', as it were. We shall stick with the Lineweaver–Burk plot, however, because it is the most widely used.

SAQ 10 Using a Hanes plot, determine K_m and V from the data given in Table 4. In what ways is this plot more satisfactory than that obtained from the Lineweaver–Burk equation? Using this value of V and the concentration of the enzyme given in Table 4, determine the rate constant k_R.

4.2 Regulation of the kinetic characteristics of an enzyme-catalysed reaction: an aside

Just as a busy traffic system needs control points, so too does an orderly flow through metabolic pathways depend on regulatory mechanisms. In the previous Section, you saw that the actual rate of an enzyme-catalysed reaction depends linearly on the concentration of the enzyme. For any fixed amount of enzyme, the rate also depends on:

- the concentration(s) of the substrate(s); and

- the kinetic characteristics of the enzyme itself — that is, K_m (the Michaelis constant) and V (the limiting rate).

This offers several different possible control mechanisms. Here, we look briefly at just one of the strategies employed by living things — the production of enzymes with *different* kinetic characteristics to catalyse the same reaction. This should give you a feel for the physiological significance of the quantities V and K_m.

If an enzyme has a small value of K_m, the condition $[S]_0 \gg K_m$ will be achieved when the concentration of substrate is relatively small. Hence the limiting rate is approached at quite low substrate concentrations: see, for example, Figure 22, curve (a). On the other hand, if K_m is large, a high concentration of substrate is required for $[S]_0 \gg K_m$, so that the enzyme only reaches its limiting rate when the concentration of substrate is large (curve (b) in Figure 22). We conclude that *the Michaelis constant governs the range of substrate concentrations over which the initial rate approaches its limiting value.*

- How do you think altering the value of V affects the range of substrate concentrations over which the rate reaches its limiting value?

- Altering V only changes the limiting rate (compare curves (c) and (a)); it has *no* effect on the relative sizes of $[S]_0$ and K_m, and hence no effect on the concentration of substrate required to saturate the enzyme.

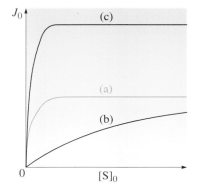

Figure 22 The effect of changing K_m (curves (a) and (b)) and V (curves (a) and (c)) on the plot of J_0 against $[S]_0$ for an enzyme-catalysed reaction obeying the Michaelis–Menten equation.

The maximum rate of an enzyme-catalysed reaction can also be increased by simply increasing the amount of enzyme present: this is indeed one of the ways in which cells regulate enzyme activity. However, this does not affect K_m; that is, it does not alter the response characteristics of the enzyme. But if the physiological situation changes, it may become desirable to alter the response characteristics.

Take, for example, the *phosphorylation* of glucose to glucose 6-phosphate by ATP (adenosine triphosphate):

glucose + adenosine triphosphate $\longrightarrow$ glucose 6-phosphate + adenosine diphosphate   (69)

In muscle and tissues with similar metabolic needs, this reaction forms the first step of the energy-yielding glycolytic pathway — sometimes referred to as glycolysis (Figure 12) — in which glucose is converted to lactate. On the other hand, the liver has a different role and therefore handles glucose differently. Among other things, the liver has to respond after a meal to the tide of glucose coming in from the intestine via the portal vein. This concentration is very much higher than that reaching the muscle. The first reaction, however, is still a conversion to glucose 6-phosphate, as in muscle. This is an example of the body producing *two* different enzymes to catalyse the same reaction — *glucokinase* (found in the liver) and *hexokinase* (found in muscle). In other words, there are separate genes coding for two different proteins, both of which catalyse the phosphorylation, but which nevertheless *differ* in their kinetic properties. When different enzymes catalyse the same reaction within a single species, they are known as **isoenzymes**. Because these two enzymes operate under different conditions, they have very different K_m values for glucose: glucokinase has a K_m of 10^{-2} mol dm^{-3}, whereas hexokinase has a K_m of 10^{-5} mol dm^{-3}.

■ Which of these two enzymes becomes saturated at the lower glucose concentration?

■ As we have seen, the Michaelis constant defines the range of substrate concentration over which the enzyme approaches the limiting rate. Since K_m is smaller for hexokinase than glucokinase, it will be saturated first, as shown in Figure 23.

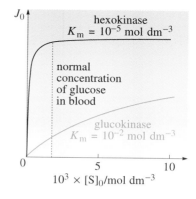

Figure 23 Comparison of the kinetic behaviour of hexokinase and glucokinase.

It is not surprising that an enzyme like glucokinase with a comparatively large K_m is found in the liver. An enzyme that was incapable of responding to the high levels of glucose likely to be provided from time to time in the portal blood would be of little value. An enzyme with a small K_m would *always* be working flat out over these glucose concentrations, and would thus be insensitive to changes in the incoming glucose concentration. On the other hand, an enzyme with a high K_m would be almost useless in muscle, which, like all tissues other than the intestine and liver, has to obtain its glucose from the main blood circulation. Under these conditions, such an enzyme would never reach more than a tiny fraction of its potential maximum rate.

4.3 Summary of Sections 4–4.2

1   Enzymes are very efficient, highly selective catalysts.

2   Enzymes are extremely large naturally occurring molecules made up of amino acid units. There are twenty common amino acids, all but one of which are chiral. The polypeptide chain in an enzyme folds up to give a precise shape which contains an active site. This active site provides the perfect environment for the reactant(s).

3   A series of enzymes will select just one of a large number of reaction pathways. This is called a metabolic pathway.

4   Enzyme kinetic studies usually involve measurement of the initial rates. *In vitro*, the chemical rate equation of enzyme-catalysed reactions involving one substrate is first order in the enzyme, and of variable order in the substrate. *At fixed enzyme concentration*, the variable order in the substrate can be expressed mathematically by the Michaelis–Menten equation (equation 56):

$$J_0 = \frac{V[S]_0}{K_m + [S]_0} \qquad (56)$$

V is the limiting rate at high substrate concentrations, and K_m (the Michaelis constant) is equivalent to the concentration of substrate that gives half the maximum rate. The value of K_m governs the range of substrate concentrations over which the initial rate approaches its limiting value.

5   Equation 56 can be manipulated to give a number of linear forms which allow K_m and *V* to be determined graphically.

6   Living organisms often produce more than one protein catalyst for the same reaction. These isoenzymes usually have distinctive kinetic properties (for example, different values of K_m). They also often have characteristic distributions within the organism, which may provide a clue to their separate biological functions. They are needed because any given enzyme-catalysed reaction may form part of more than one metabolic pathway or process, and the separate roles demand separate control characteristics.

4.4 The mechanism of enzyme action

In this Section we shall examine the *mechanism* of enzyme-catalysed reactions. As you would expect, the experimental rate equation tells us a lot about the possible pathways.

▣ Do you think that enzyme-catalysed reactions that have experimental rate equations similar to equation 55, such as the conversion of malate into fumarate, will involve composite mechanisms?

$$J_0 = \frac{k_R [E]_0 [S]_0}{K_m + [S]_0} \tag{55}$$

▣ Following the strategy outlined in Block 3, the question we should be asking is: can such reactions proceed via a single elementary step and still have the rate equation 55? The answer is, of course, *no*. Elementary reactions have simple first- or second-order (or rarely, third-order) rate equations only; equation 55 is *far* too complex. In other words, enzyme-catalysed reactions must have composite mechanisms.

Having established the need for a composite mechanism, the next step is to find the simplest mechanism that is consistent with the experimental rate equation 55. Most one-substrate reactions, such as the one we met in Section 4.1, can be written in the following general form:

$$\text{substrate (S)} \xrightarrow{\ \ E\ \ } \text{products (P)} \tag{70}$$

The simplest composite mechanism that can be proposed for such a conversion is a two-step pathway involving one intermediate, a **substrate–enzyme complex**, S.E:

$$E + S \underset{k_{-1}}{\overset{k_1}{\rightleftharpoons}} S.E \tag{71}$$

$$S.E \xrightarrow{\ k_2\ } E + P \tag{72}$$

Because most enzyme-catalysed reactions exhibit time-independent stoichiometry, as did our previous example, the intermediate complex must not build up to significant proportions. (Recall Block 3, Section 3.2.)

▣ How can we test the validity of this mechanism?

▣ An obvious test is to check whether this mechanism predicts a chemical rate equation in agreement with its experimental counterpart. (In this case we shall employ the full form, equation 55.)

▣ Write down the theoretical rate equations for the product P and the intermediate S.E, based on the elementary steps of the proposed mechanism.

$$\frac{d[P]}{dt} = k_2 [S.E] \tag{73}$$

$$\frac{d[S.E]}{dt} = k_1 [S][E] - k_{-1} [S.E] - k_2 [S.E] \tag{74}$$

Because S.E does not build up to significant proportions, we can simplify the analysis of the theoretical rate equations by applying the steady-state approximation.

▣ Apply the steady-state approximation to the intermediate S.E, and thus obtain an expression for its concentration in terms of the concentrations of the substrate and the enzyme.

◻ The steady-state approximation allows us to put both sides of equation 74 equal to zero; that is,

$$0 = k_1[S][E] - k_{-1}[S.E] - k_2[S.E] \tag{75}$$

Collecting terms in $[S.E]$ gives

$$0 = k_1[S][E] - (k_{-1} + k_2)[S.E] \tag{76}$$

which can be rewritten as

$$(k_{-1} + k_2)[S.E] = k_1[S][E] \tag{77}$$

so $[S.E] = \dfrac{k_1}{(k_{-1} + k_2)}[S][E]$ $\tag{78}$

Combining equations 73 and 78 gives us the theoretical rate equation for the product in terms of the concentrations of the substrate and the enzyme:

$$\frac{d[P]}{dt} = \frac{k_1 k_2}{(k_{-1} + k_2)}[S][E] \tag{79}$$

◻ From the definition of J_0 for the overall reaction, write down the chemical rate equation, in terms of the *initial rate*, as predicted by this mechanism.

◻ $J_0 = \dfrac{d[P]_0}{dt} = \left(\dfrac{k_1 k_2}{k_{-1} + k_2}\right)[S]_0[E]$ $\tag{80}$

In answering this question you may have tried to substitute $[E]_0$ for $[E]$ in equation 80 — it seems an obvious thing to do — to give the expression

$$J_0 = \left(\frac{k_1 k_2}{k_{-1} + k_2}\right)[S]_0[E]_0 \tag{81}$$

However, this would have been wrong; don't worry, we wouldn't expect you to have realized this. The problem arises because there is a subtle difference between the definitions of the two terms $[E]$ and $[E]_0$. The term $[E]$ that we have just used in our derivation of the chemical rate equation is the concentration of the enzyme *that is free to catalyse the reaction,* whereas $[E]_0$ is the concentration of the enzyme *that was added to the reaction mixture* at the beginning of the reaction; as you will see, these are *not* necessarily the same. Enzymes are very efficient catalysts and, as the data in Table 3 (Section 4.1) show, only a very small concentration of the enzyme is required to catalyse the reaction. As it turns out, this concentration is so small that the concentration of the enzyme that is 'tied up' in the substrate–enzyme complex, $[S.E]$, is no longer insignificant compared with the total enzyme concentration; this means that the concentration of free enzyme is *not* equal to that added at the start of the reaction.

◻ How are $[E]$, $[S.E]$ and $[E]_0$ related?

◻ The total concentration of enzyme added to the reaction mixture will equal the sum of the concentration of the free enzyme plus that tied up in the substrate–enzyme complex:

$$[E]_0 = [E] + [S.E] \tag{82}$$

Because we cannot substitute $[E]_0$ for $[E]$ directly, it is useless trying to compare the chemical rate equation 80 with its experimental counterpart, equation 55. Clearly, we need to obtain another expression containing $[E]_0$. We could modify equation 80, but this is quite complicated. An easier approach is to go back a few steps in our derivation, and obtain an expression for $[S.E]$ in terms of $[E]_0$ rather than $[E]$.

To this end, we start by rewriting equation 82 as follows:

$$[E] = [E]_0 - [S.E] \tag{83}$$

and combine this expression with equation 77 to give

$$(k_{-1} + k_2)[S.E] = k_1[S][E]_0 - k_1[S][S.E] \tag{84}$$

Now we collect terms in [S.E], as we did before:

$$(k_{-1} + k_2 + k_1[S])[S.E] = k_1[S][E]_0 \tag{85}$$

and divide both sides by k_1 (you will see why in a moment):

$$\left\{ \left(\frac{k_{-1} + k_2}{k_1} \right) + [S] \right\} [S.E] = [S][E]_0 \tag{86}$$

to give the following expression for [S.E]:

$$[S.E] = \frac{[S][E]_0}{\left\{ \left(\dfrac{k_{-1} + k_2}{k_1} \right) + [S] \right\}} \tag{87}$$

By combining equations 73 and 87, we now obtain the theoretical rate equation for the product in terms of [S] and $[E]_0$:

$$\frac{d[P]}{dt} = \frac{k_2[S][E]_0}{\left\{ \left(\dfrac{k_{-1} + k_2}{k_1} \right) + [S] \right\}} \tag{88}$$

From the definition of J_0 for the overall reaction, we can write down the chemical rate equation, in terms of the initial rate, that is predicted by this mechanism:

$$J_0 = \frac{d[P]_0}{dt} = \frac{k_2[S]_0[E]_0}{\left\{ \left(\dfrac{k_{-1} + k_2}{k_1} \right) + [S]_0 \right\}} \tag{89}$$

This chemical rate equation is expressed in terms of $[E]_0$, and *can* be compared with its experimental counterpart:

$$J_0 = \frac{k_R[E]_0[S]_0}{K_m + [S]_0} \tag{55}$$

■   How does it compare? How would you interpret the experimental terms k_R and K_m?

▨   Equations 55 and 89 are of a similar form, with k_R equivalent to k_2 and K_m equivalent to $(k_{-1} + k_2)/k_1$.

This mechanism *is* consistent with the experimental rate equation (which should be no surprise to you; otherwise why choose it?) and, until evidence is provided to the contrary, we shall assume it operates for *all* one-substrate enzyme reactions.

Having formulated a reasonable mechanism, we can now acquire some insight into the strange behaviour of the initial rate when the concentration of the substrate is changed.

■   What happens to equation 89 at high concentrations of the substrate? What does this behaviour signify?

■ At high substrate concentrations, $[S]_0$ dominates the denominator, and equation 89 reduces to

$$J_0 = k_2[E]_0 \tag{90}$$

Comparison of this expression with equation 73 suggests that under these conditions the second step in the mechanism (equation 72) is rate limiting, virtually all the enzyme being tied up in the substrate–enzyme complex (that is, $[S.E] \approx [E]_0$).

This corresponds to the horizontal portion of the curve shown in Figure 19 (Section 4.1): as we noted there, the enzyme is said to be **saturated** with the substrate. Almost as soon as a substrate–enzyme complex decomposes to product and free enzyme, the latter reacts with another substrate molecule to form more of the complex. Adding *more* substrate then has no effect on the rate because the rate depends only on the decomposition of the complex in the second step.

■ What happens to equation 89 at low substrate concentrations?

■ At low substrate concentrations the term $(k_{-1} + k_2)/k_1$ dominates the denominator, and equation 89 reduces to

$$J_0 = \left(\frac{k_1 k_2}{k_{-1} + k_2} \right)[S]_0 [E]_0 \tag{91}$$

■ Look back at equation 80. What can you say about the concentration of the free enzyme in the region where equation 91 holds?

■ Equation 80 is the chemical rate equation predicted by this mechanism, but expressed in terms of the concentration of the *free* enzyme. Equation 91 has an identical form, and this in turn suggests that *in this region* the concentration of the free enzyme is virtually identical with the concentration of enzyme that was added to the initial reaction mixture.

This is indeed the case. At low substrate concentrations, the concentration of the substrate–enzyme complex is so small that it is insignificant, even by comparison with the *total* enzyme concentration.

In summarizing, it is useful to remind ourselves of the mechanism:

$$E + S \underset{k_{-1}}{\overset{k_1}{\rightleftharpoons}} S.E \tag{71}$$

$$S.E \xrightarrow{k_2} E + P \tag{72}$$

At low substrate concentrations, virtually all the enzyme is in the free state, with little tied up in the complex. In this region the limiting case represented by equation 91 is valid: the reaction is first order with respect to the substrate. As the concentration of the substrate is increased, the concentration of the complex also increases, until at high substrate concentrations the enzyme is virtually saturated with the substrate. In this region the limiting case represented by equation 90 is valid: the reaction is zeroth order with respect to the substrate. In the changeover region, neither limiting case of equation 89 is valid, because $(k_{-1} + k_2)/k_1$ and $[S]_0$ are of a similar size.

SAQ 11   If the first step in the mechanism shown in equations 71 and 72 is a *rapidly established pre-equilibrium*, what does the term $(k_{-1} + k_2)/k_1$ in equation 89 reduce to? Under these circumstances, how is K_m related to the dissociation constant, K_D, of the substrate–enzyme complex?

$$S.E = S + E \tag{92}$$

$$K_D = \frac{[S][E]}{[S.E]} \tag{93}$$

It is important that you work through SAQ 12 before moving on. It invites you to analyse an alternative mechanism for an enzyme-catalysed reaction as a prelude to the discussion in the next Section.

SAQ 12 The mechanism shown in equations 71 and 72 was first proposed by Victor Henri in 1902. He also proposed a second mechanism, which also agreed with his experimental results:

$$E + S \underset{k_{-1}}{\overset{k_1}{\rightleftharpoons}} S.E \tag{71}$$

$$E + S \xrightarrow{k_2} E + P \tag{94}$$

in which the complex S.E is a '**nuisance complex**', which doesn't lead anywhere but merely ties up some of the enzyme. According to this mechanism, the product is formed in a completely unrelated step, equation 94.

(a) Assuming that the first step is a rapidly established pre-equilibrium, write an expression for [S.E] in terms of [E] and [S].

(b) Use this equation to substitute for [S.E] in the equation for the initial concentration of enzyme, $[E]_0$ (for example, equation 82), and thus derive an expression for [E] in terms of [S] and $[E]_0$.

(c) Use this expression to derive the chemical rate equation for J_0, the initial rate of conversion of S into P, in terms of $[S]_0$ and $[E]_0$. Compare your expression with equation 56. How would you interpret the experimental parameters V and K_m in this case?

4.5 Transient kinetics

As you should have found in answering SAQ 12, the second mechanism that Henri proposed also generates a chemical rate equation that is consistent with the experimental rate equation. We thus have two possible mechanisms with identical kinetic forms, and, as yet, we are not able to decide experimentally which is the more likely.

Mechanism A

$$E + S \underset{k_{-1}}{\overset{k_1}{\rightleftharpoons}} S.E \tag{71}$$

$$S.E \xrightarrow{k_2} E + P \tag{72}$$

Mechanism B

$$E + S \underset{k_{-1}}{\overset{k_1}{\rightleftharpoons}} S.E \tag{71}$$

$$E + S \xrightarrow{k_2} E + P \tag{94}$$

■ What is the difference between the roles of the substrate–enzyme complex in these two mechanisms?

▨ In Mechanism B, the substrate–enzyme complex, S.E, *is not* part of the catalysed route, which only involves the second step, equation 94. In Mechanism A, the substrate–enzyme complex is an intermediate in the catalysed pathway.

Although these two mechanisms have indistinguishable kinetics in the steady-state situation, they do differ in the way they *attain* this steady state. To show this, let us first look at how each mechanism predicts the steady state will be approached, and then go on to examine a real example that allows us to distinguish between the two pathways.

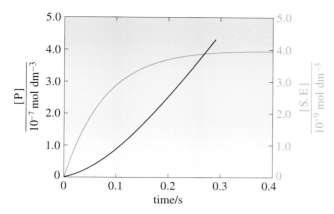

Figure 24 A computer simulation of an enzyme-catalysed reaction proceeding via Mechanism A, showing the change in the concentrations of the product P (black curve) and the substrate–enzyme complex, S.E (blue curve). Note the different scales for the two concentrations.

Figure 24 shows the results of a computer simulation of a typical enzyme-catalysed reaction proceeding via Mechanism A. We have assumed typical values of k_1, k_{-1} and k_2, and determined how the concentrations of the product and the substrate–enzyme complex vary during the very early stages of reaction. There is a very quick build-up in the concentration of the substrate–enzyme complex until a plateau is reached, where the rate of its formation is balanced by the rate of its loss. This is the steady-state region, and the concentration of the complex decreases only *slowly* as the substrate is used up. Clearly, this validates the use of the steady-state approximation, because for most of the reaction $d[S.E]/dt$ is *very* small.

Now look at how the concentration of the product changes in Figure 24. At the start there seems to be a 'lag' in the rate of product formation, before the reaction really gets going. This is because according to Mechanism A the substrate and enzyme must first be converted into the complex *before* the product is formed. As we saw, it takes a little while for the concentration of the substrate–enzyme complex to reach its optimum value.

Figure 25 shows the results of an equivalent simulation for Mechanism B. The variation in the concentration of the substrate–enzyme complex is very similar to that observed for Mechanism A, so we have not shown it in Figure 25. However, the variation in the concentration of the product with time shows a 'burst', not a 'lag'. This is because the rate of product formation would be *largest* at the beginning of the reaction, when more of the enzyme exists in the free state and can thus take part in the catalysed process (reaction 94). As the concentration of the 'nuisance complex' increases, that of the free enzyme decreases, as does the rate of product formation.

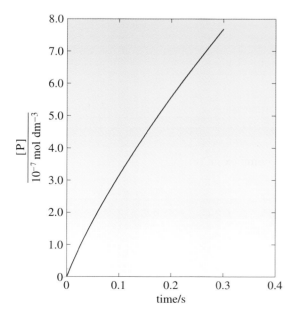

Figure 25 A computer simulation of the enzyme-catalysed reaction of Figure 24 proceeding via Mechanism B, showing the change in the concentration of the product.

So, by studying the way in which the enzyme-catalysed reaction attains the steady state — that is, by conducting a **pre-steady-state study** — we should be able to distinguish between the two mechanisms.

The hydrolysis of certain peptide bonds in proteins is catalysed by *ficin*, an enzyme found in figs:

$$H_2O + R\!-\!\underset{\underset{O}{\|}}{C}\!-\!NHR' \xrightarrow{\text{ficin}} R\!-\!\underset{\underset{O}{\|}}{C}\!-\!OH + NH_2R' \tag{95}$$

Figure 26 shows how the concentration of the product varies with time in a pre-steady-state study of the ficin-catalysed hydrolysis of a simple amide. Unfortunately, as a result of experimental difficulties, it was not possible to determine absolute values for the concentrations, and these are therefore quoted in arbitrary units. Nevertheless, the plot does give an idea of the manner in which the concentration of the product changes *at the start of the reaction.*

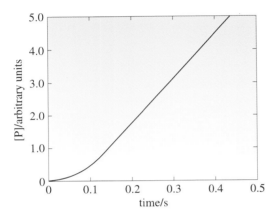

Figure 26   The variation in the concentration of the product in a pre-steady-state study of the hydrolysis of a simple amide catalysed by ficin.

■ Which mechanism, A or B, do you think is operating in this system?

■ Figure 26 clearly shows a lag in the rate of product formation at the beginning of the reaction. This suggests that the two-step mechanism, A, is in operation.

As well as steady-state studies, such as those described in Section 4.1 (where the initial concentration of the substrate is in large excess over the total enzyme concentration), and the pre-steady-state studies we have just discussed, biochemists also use **non-steady-state kinetics**. This involves employing concentrations of the enzyme that are similar to — or greater than — that of the substrate. Pre-steady-state and non-steady-state kinetics are collectively known as **transient kinetics**. Such studies yield valuable information on how the enzyme works — information that is *not* available from the corresponding steady-state studies.

Whereas the mechanism shown by equations 71 and 72 is consistent with steady-state kinetics, evidence from other experiments demonstrates that the situation is more complicated than this. While the substrate is bound to the enzyme, it generally undergoes a chemical change, *before* being released as the product. Our mechanism needs to be modified to take account of this, one obvious possibility being as follows:

$$E + S \rightleftharpoons S.E \longrightarrow S'.E \longrightarrow E + P \tag{96}$$

Application of the steady-state approximation to the reaction intermediates S.E and S'.E still gives a chemical rate equation of the same form as the Michaelis–Menten equation, but the constants V and K_m have a slightly different significance. Clearly, if the chemical rate equations predicted by the two-step mechanism (equations 71 and 72) and the three-step mechanism (equation 96) are of *exactly* the same form, steady-state studies cannot provide us with any information about the *number* of intermediates along the pathway.

■ In steady-state kinetics, the concentration of the enzyme is *very* small compared with that of the substrate. What does this tell us about the number of times a molecule of enzyme goes through the catalytic cycle during the course of the overall reaction? What does this imply about the speed of steps, such as the conversion of S.E into S'.E, in equation 96?

■ Because the reactant is in great excess, a molecule of enzyme must go through the catalytic cycle very many times during the reaction in order to convert all the reactant to product. This means that steps like the conversion of S.E into S'.E must occur very quickly in comparison with the time-scale of the overall reaction.

Although it is impossible to pick out the individual steps of the catalytic cycle using steady-state kinetics, if the concentration of the enzyme is similar to or greater than that of the substrate, we can investigate a *single* trip round the cycle. With luck we may be able to detect any intermediates or strange kinetic behaviour, and thus, as we did in Block 3, develop a mechanism that accounts for such behaviour. Transient kinetic experiments require measurements to be made in the time range from 10^{-7} to 1 second. Clearly, unless special techniques are employed, the reaction will be over by the time the two reactants are completely mixed together. **Stopped-flow spectro-photometry** is a technique that allows the mixing of reactants in a fraction of a millisecond; it is particularly useful in this context because *most* enzyme-catalysed processes have half-lives greater than this value. You will have a chance to measure the rates of fast reactions using the stopped-flow technique at the Residential School.

The use of transient kinetics has been made possible through the development of apparatus for measuring rapid reactions and of techniques for isolating large amounts of pure enzyme. Such techniques have allowed scientists to probe the manner in which enzymes catalyse reactions, and we are slowly learning the secrets of the enzymes' success. As a result, scientists are learning to design catalysts that can mimic enzyme actions in the laboratory. The study of the mechanism of enzyme-catalysed actions *in vitro* has also taught us much about how chemical reactions are carried out in the *living* organism (*in vivo*), and about how these processes are controlled to meet the various demands placed on the organism.

4.6 Summary of Sections 4.4 and 4.5

1   The steady-state kinetic behaviour of enzyme-catalysed reactions can be modelled by a two-step mechanism involving a substrate–enzyme complex. Application of the steady-state approximation to this reaction intermediate gives a rate equation, which, after allowance for the concentration of the enzyme tied up in the substrate–enzyme complex, gives a chemical rate equation of the same form as its experimental counterpart (equation 56). There are two limiting forms of this rate equation:

(a)   At high substrate concentrations, the enzyme is saturated with the substrate, and the rate depends solely on the rate of decomposition of the substrate–enzyme complex.

(b)   At low substrate concentrations, the enzyme is mainly in the 'free state', and the rate depends on the concentration of both the substrate and the enzyme.

2   There is a second possible mechanism involving the equilibrium formation of a 'nuisance complex' and a one-step catalytic process, which also gives a theoretical rate equation similar to equation 56. Analysis of the pre-steady-state kinetics allows a distinction to be made between the two mechanisms, and it is found that the former mechanism is more likely.

3   Non-steady-state kinetics, in which the concentration of the enzyme is similar to or greater than that of the substrate, can be used to obtain valuable information on the number of intermediates along the reaction pathway. Such information is not readily available from steady-state studies.

The important equations developed in Section 4 are collected together in Box 2.

Box 2   Summary of enzyme kinetics

The experimental rate equation of many enzyme-catalysed reactions takes the form:

$$J_0 = \frac{k_R [E]_0 [S]_0}{K_m + [S]_0} \tag{55}$$

Because $[E]_0$ is often kept constant, this reduces to the Michaelis–Menten equation:

$$J_0 = \frac{V [S]_0}{K_m + [S]_0} \tag{56}$$

where V is the limiting rate at high substrate concentrations ($V = k_R [E]_0$), and K_m is the Michaelis constant, which is equal to the concentration of substrate that gives half the limiting rate.

Equation 56 can be expressed in another form, known as the Lineweaver–Burk equation:

$$\frac{1}{J_0} = \frac{K_m}{V} \left(\frac{1}{[S]_0} \right) + \frac{1}{V} \tag{66}$$

The simplest mechanism of enzyme action is of the form:

$$E + S \underset{k_{-1}}{\overset{k_1}{\rightleftharpoons}} S.E \tag{71}$$

$$S.E \overset{k_2}{\longrightarrow} E + P \tag{72}$$

Application of the steady-state approximation to the reaction intermediate S.E gives the chemical rate equation predicted by this mechanism:

$$J_0 = \frac{k_2 [S]_0 [E]_0}{\left\{ \left(\dfrac{k_{-1} + k_2}{k_1} \right) + [S]_0 \right\}} \tag{89}$$

OBJECTIVES FOR BLOCK 4

Now that you have completed Block 4, you should be able to do the following things:

1   Recognize valid definitions of, and use in a correct context, the terms, concepts and principles printed in bold type in the text and collected in the following Table.

List of scientific terms, concepts and principles used in Block 4

Term	Page No
acid and base catalysis	11
active site	27
amino acid	26
catalyst	7
catalytic cycle	11
coordination compound	15
coordination number	19
denaturation	29
Eadie plot	33
enzyme	25
enzyme saturation	39
Hanes plot	33
heterogeneous catalysis	6
homogeneous catalysis	6
hydroformylation (oxo process)	19
in vitro	29
in vivo	29
isoenzyme	35
ligand	16
limiting rate (V)	31
Lineweaver–Burk plot	33
metabolic pathway	28
Michaelis constant (K_m)	30
Michaelis–Menten equation	30
non-participative ligand	19
non-steady-state kinetics	42
nuisance complex	40
oxidation numbers (states)	16
oxidative addition	19
participative ligand	19
poisoning (of catalysts)	22
pre-steady-state study	42
reductive elimination	19
selectivity	5
steric effect	21
stopped-flow spectrophotometry	43
substrate	29
substrate–enzyme complex	36
transient kinetics	42
transition metal catalysis	15
transition metal complex	15

2   Discuss the effect of a catalyst on:

   (a)   the stoichiometry and Gibbs free energy change of a reaction; (SAQs 2 and 4)

   (b)   the kinetics and mechanism of a reaction; (SAQs 1, 2 and 4)

   (c)   the activation energy and the *A*-factor of a reaction.

3   Outline the main features of a catalytic cycle. (SAQs 3, 4 and 5)

4   Discuss why transition metal complexes make good catalysts. (SAQs 6 and 7)

5   Discuss the similarities and differences between homogeneous, heterogeneous, and enzyme catalysis.

6   Outline some of the implications of using transition metal homogeneous catalysts in industrial processes. (SAQ 7)

7   Describe the structure of enzymes and the typical features of enzyme-catalysed reactions.

8   Discuss the form of the steady-state chemical rate equation of enzyme-catalysed reactions and the various methods by which K_m and V may be determined. (SAQs 8, 9 and 10)

9   Interpret the values of K_m for different isoenzymes in a given organism in terms of the range of substrate concentrations over which the enzyme approaches its limiting rate.

10   Outline a mechanism that accounts for the steady-state kinetic behaviour of enzyme-catalysed reactions, and use it to interpret the meaning of the limiting cases of the rate equation. (SAQs 11 and 12)

11   Use transient kinetic data to establish the mechanism of enzyme-catalysed reactions.

SAQ ANSWERS AND COMMENTS

SAQ I (Objective 2)

According to the overall stoichiometric equation (equation 4), the rate of reaction is given by:

$$J = -\frac{1}{2}\frac{d[H_2O_2]}{dt} = \frac{1}{2}\frac{d[H_2O]}{dt} = \frac{d[O_2]}{dt} \qquad (97)$$

Using the strategy outlined in Block 3, the next step is to write down the theoretical rate equation for the reactant (H_2O_2) or for one of the products (H_2O and O_2). In this case, O_2 is only involved in the second step of the mechanism, so it simplifies matters to choose this species; that is:

$$\frac{d[O_2]}{dt} = k_2[IO^-][H_2O_2] \qquad (98)$$

Applying the steady-state approximation to the *reactive* intermediate (IO^-) implies that $d[IO^-]/dt = 0$; that is

$$\frac{d[IO^-]}{dt} = 0 = k_1[H_2O_2][I^-] - k_2[IO^-][H_2O_2] \qquad (99)$$

which gives the following expression for $[IO^-]$:

$$[IO^-] = \frac{k_1[H_2O_2][I^-]}{k_2[H_2O_2]} = \frac{k_1[I^-]}{k_2} \qquad (100)$$

Substituting this expression for [IO⁻] into equation 98 gives:

$$\frac{d[O_2]}{dt} = \frac{k_1 k_2 [H_2O_2][I^-]}{k_2} = k_1[H_2O_2][I^-] \qquad (101)$$

Combining equations 97 and 101 gives the chemical rate equation predicted by this mechanism:

$$J = \frac{d[O_2]}{dt} = k_1[H_2O_2][I^-] \qquad (102)$$

This is in agreement with its experimental counterpart, equation 6.

We did not ask you to do this, but it is worth noting that the theoretical rate equation for the *catalyst* (I⁻), based on the elementary steps of the mechanism is given by:

$$\frac{d[I^-]}{dt} = -k_1[H_2O_2][I^-] + k_2[IO^-][H_2O_2] \qquad (103)$$

Substituting in the expression for [IO⁻] from equation 100, this becomes:

$$\frac{d[I^-]}{dt} = -k_1[H_2O_2][I^-] + \frac{k_1 k_2[H_2O_2][I^-]}{k_2} = 0 \qquad (104)$$

In other words, the concentration of I⁻ does *not* change during the reaction, which is in accordance with the definition of a catalyst.

SAQ 2 (Objective 2)

When substituted into equation 16, the values listed in the question give the following expression:

$$J = \{(9.2 \times 10^{-7}\,s^{-1}) + (1.7 \times 10^{-2}\,dm^3\,mol^{-1}\,s^{-1})[H^+]$$
$$+ (1.7 \times 10^{-4}\,dm^3\,mol^{-1}\,s^{-1})[OH^-]\}[C_2H_4O] \qquad (105)$$

(i) At high acidity (say pH = 1, where [H⁺] = 0.1 mol dm⁻³ and [OH⁻] = (10⁻¹⁴/10⁻¹) mol dm⁻³ = 10⁻¹³ mol dm⁻³), the middle term will be the largest by far. Hence, in this pH region, the expression reduces to:

$$J = k_A[H^+][C_2H_4O] \qquad (106)$$

(ii) In neutral solution (pH = 7, where [H⁺] = 10⁻⁷ mol dm⁻³ and [OH⁻] = 10⁻⁷ mol dm⁻³), the first term will dominate, and so in this region the expression reduces to:

$$J = k_0[C_2H_4O] \qquad (107)$$

(iii) In basic solution (say pH = 14, where [H⁺] = 10⁻¹⁴ mol dm⁻³ and [OH⁻] = (10⁻¹⁴/10⁻¹⁴) mol dm⁻³ = 1.0 mol dm⁻³), the last term will dominate, and the expression reduces to:

$$J = k_B[OH^-][C_2H_4O] \qquad (108)$$

Equation 106 indicates that at high acidity the rate depends on the concentration of hydrogen ions, but the stoichiometry of the overall reaction, given by equation 15, shows that hydrogen ions are *not* consumed in this reaction. Thus, H⁺ must be a catalyst: in this region the reaction proceeds via an *acid-catalysed* route. A similar analysis shows OH⁻ to be a catalyst in basic solution; in other words, there is also a base-catalysed pathway. There is no evidence for any catalytic process in neutral solution; in this region the reaction proceeds via an uncatalysed route.

In fact, the industrial manufacture of ethane-1,2-diol employs acid catalysis.

SAQ 3 (Objective 3)

Starting with the catalyst, Mn^{II}, which (by convention) is located at the top of the cycle, and proceeding in a clockwise fashion, the first step we meet is

$$Mn^{II} + Ce^{IV} \longrightarrow Mn^{III} + Ce^{III} \tag{110}$$

Continuing round the cycle, the next step we meet is:

$$Mn^{III} + Ce^{IV} \longrightarrow Mn^{IV} + Ce^{III} \tag{111}$$

Finally, we come to the third step, in which Mn^{IV} is converted back into Mn^{II}:

$$Mn^{IV} + Tl^{I} \longrightarrow Mn^{II} + Tl^{III} \tag{112}$$

SAQ 4 (Objectives 2 and 3)

(a)   Hydrogen ions, H^+, appear in the experimental rate equation, so the reaction will go faster in acid solution. Since H^+ does not appear in the stoichiometric equation, it must be acting as a catalyst.

(b)   Since iodide ion, I^-, is involved in the mechanism (and will thus affect the *rate* of reaction), but does *not* appear in the stoichiometric equation, it can be classed as a catalyst.

(c)   In this system, NO is acting as a catalyst. Remember that by convention the catalyst appears at the top of the cycle: proceeding in a clockwise direction, NO is first converted into NO_2, which is then converted back into NO. Meanwhile, SO_2 and $\frac{1}{2}O_2$ are converted into SO_3. We can express this mechanism in equations as:

$$NO + \tfrac{1}{2}O_2 \longrightarrow NO_2 \tag{113}$$

$$NO_2 + SO_2 \longrightarrow NO + SO_3 \tag{114}$$

SAQ 5 (Objective 3)

$P(C_6H_5)_3$ is a ligand with oxidation number 0, as is CO. Hydrogen in metal hydrides has an oxidation number of -1, as do alkyl and other hydrocarbon ligands linked by σ bonds to the metal. Since all of the Rh complexes in Figure 7 are neutral species, you should have deduced the following:

Complex	Oxidation number of Rh
10	+1
11	+1
12	+1
13	+1
14	+1
15	+1
16	+3

L is a non-participative ligand in this cycle.

SAQ 6 (Objective 4)

(a)   The participative ligands — those that take an active part in the catalytic cycle — are H, CN, $CH_2=CH-CH=CH_2$, $CH_2-CH=CH-CH_3$ and two of the Ls in **28** (where L is the $P(OC_6H_5)_3$ ligand). The other two Ls in **28** are non-participative, and do not participate directly in the catalytic cycle.

(b)   The hydrogen cyanide enters the cycle during the conversion **28 → 29** and the 1,3-butadiene enters at **29 → 30**. The product is lost when **31** is converted back to **28**.

(c)   The step **28 → 29** involves an oxidative addition. In metal hydrides, the oxidation number of hydrogen is taken as −1, the same as the cyanide ligand, so the oxidation number of the nickel increases by two, from 0 in **28** to +2 in **29**.

(d)   The coordination number of nickel changes from five to four when the complex **30** is converted to **31**.

SAQ 7 (Objectives 4 and 6)

Statements (a), (c) and (e) are true. Statement (b) is false: it is a feature of transition metal ions that they form complexes with a wide range of ligands. Statement (d) is false: it is easier to remove the catalyst from the reaction mixture in heterogeneous catalysis than in its homogeneous counterpart.

SAQ 8 (Objective 8)

The plot of J_0 against $[S]_0$ obtained using the data in Table 4 is shown in Figure 27. Since V has a value of 39×10^{-8} mol dm^{-3} s^{-1}, $V/2$ will have a value of 19.5×10^{-8} mol dm^{-3} s^{-1}. From the graph, the concentration of substrate that corresponds to an initial rate of 19.5×10^{-8} mol dm^{-3} s^{-1} is about 1.6×10^{-3} mol dm^{-3}. This, therefore, is the value of K_m.

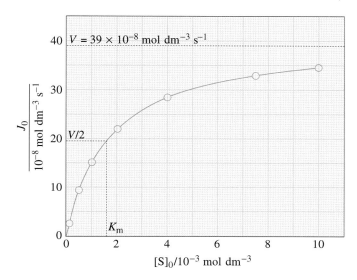

Figure 27   A plot of J_0 versus $[S]_0$ for the enzyme-catalysed reaction 52 using the data in Table 4 ($[E]_0 = 1.6 \times 10^{-9}$ mol dm^{-3} throughout).

SAQ 9 (Objective 8)

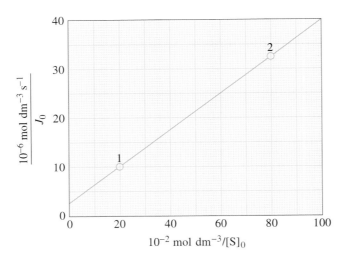

Figure 28   A Lineweaver–Burk plot of $1/J_0$ against $1/[S]_0$ using the data in Table 4.

The Lineweaver–Burk plot in Figure 21 is repeated here as Figure 28. To work out the slope of a plot that involves reciprocal quantities, it's a good plan to go back to the general expression:

$$\text{slope} = \frac{y_2 - y_1}{x_2 - x_1}$$

where, in this case, $x = 1/[S]_0$ and $y = 1/J_0$. The important thing to bear in mind is that each point on the line in Figure 28 gives you the *numerical values* of the two quantities that are used to label the axes. Thus, for example, at the point labelled 1 in Figure 28:

$$\frac{10^{-2} \text{ mol dm}^{-3}}{[S]_0} = 20 \ \text{ and } \frac{10^{-6} \text{ mol dm}^{-3} \text{ s}^{-1}}{J_0} = 10$$

so

$$x_1 = \frac{1}{[S]_0} = 20 \times 10^2 \text{ mol}^{-1} \text{ dm}^3 \ \text{ and } \ y_1 = \frac{1}{J_0} = 10 \times 10^6 \text{ mol}^{-1} \text{ dm}^3 \text{ s}$$

Using the same reasoning for the point labelled 2 yields the values:

$$x_2 = 80 \times 10^2 \text{ mol}^{-1} \text{ dm}^3; \ y_2 = 32.5 \times 10^6 \text{ mol}^{-1} \text{ dm}^3 \text{ s}$$

so the slope $= \dfrac{K_m}{V} = \dfrac{(32.5 - 10) \times 10^6 \ \text{mol}^{-1} \text{ dm}^3 \text{ s}}{(80 - 20) \times 10^2 \ \text{mol}^{-1} \text{ dm}^3}$

$$= 3.75 \times 10^3 \text{ s}$$

Similarly, the value of $1/V$ is the value of $1/J_0$ at the intercept on the y axis, so

$$\frac{1}{V} = 2.5 \times 10^6 \text{ mol}^{-1} \text{ dm}^3 \text{ s}$$

Hence

$$V = 40.0 \times 10^{-8} \text{ mol dm}^{-3} \text{ s}^{-1}$$

Since the slope $= K_m/V$ and the intercept $= 1/V$

$$\frac{\text{slope}}{\text{intercept}} = \frac{K_m/V}{1/V} = K_m = \frac{3.75 \times 10^3 \text{ s}}{2.5 \times 10^6 \ \text{mol}^{-1} \text{ dm}^3 \text{ s}}$$

$$= 1.5 \times 10^{-3} \text{ mol dm}^{-3}$$

The values of V and K_m are in good agreement with those we obtained in the text.

SAQ 10 (Objective 8)

First calculate values of $[S]_0/J_0$ for the values of $[S]_0$ in Table 4. These data (Table 5) can then be used to obtain the graph shown in Figure 29: clearly, this is a straight line. We obtained a gradient of $2.48 \times 10^6 \, \text{mol}^{-1} \, \text{dm}^3 \, \text{s}$: if your value is *very* different, check that you have handled the units and powers of ten correctly. Since the gradient is equal to $1/V$, $V = 40.3 \times 10^{-8} \, \text{mol dm}^{-3} \, \text{s}^{-1}$. The calculated value of the intercept, K_m/V, is $4.2 \times 10^3 \, \text{s}$; thus

$$K_m = (K_m/V) \times V$$

$$= (40.3 \times 10^{-8} \, \text{mol dm}^{-3} \, \text{s}^{-1}) \times (4.2 \times 10^3 \, \text{s}) = 1.69 \times 10^{-3} \, \text{mol dm}^{-3}$$

Table 5 Values of $[S]_0/J_0$ obtained from the data in Table 4

$\dfrac{[S]_0}{10^{-3} \, \text{mol dm}^{-3}}$	$\dfrac{[S]_0 / J_0}{10^3 \, \text{s}}$
0.1	4.0
0.5	5.3
1.0	6.5
2.0	9.1
4.0	14.1
7.5	22.7
10.0	29.0

Figure 29 A Hanes plot of $[S]_0/J_0$ against $[S]_0$ from the data in Table 5.

This plot is more satisfactory than that obtained from the Lineweaver–Burk equation because the points are more evenly spread across the graph, and thus those for low concentrations of substrate are not given undue weighting.

The value of k_R can be found from equation 57:

$$V = k_R[E]_0 \tag{57}$$

Thus

$$k_R = V/[E]_0$$

From the Table 4 heading, $[E]_0 = 1.6 \times 10^{-9} \, \text{mol dm}^{-3}$, and we have just calculated V to be $40.3 \times 10^{-8} \, \text{mol dm}^{-3} \, \text{s}^{-1}$. Thus, k_R is equal to $2.5 \times 10^2 \, \text{s}^{-1}$.

SAQ 11 (Objective 10)

As you may remember from Block 3, *if* the first step is an established pre-equilibrium, then the rate of the back reaction of the first step must be much greater than the rate of the second step. Molecules of the enzyme and the substrate are converted into the complex, S.E, and back again, many times before going on to give the product. This means that k_{-1} is much greater than k_2, and so $(k_{-1} + k_2)/k_1$ (which is equivalent to K_m) reduces to k_{-1}/k_1. This is equal to the dissociation constant, K_D, of the substrate–enzyme complex, as represented by equation 92, the *reverse* of equation 71. Thus

$$\text{S.E} \underset{k_1}{\overset{k_{-1}}{\rightleftarrows}} \text{E} + \text{S} \tag{115}$$

$$K_D = \frac{k_{-1}}{k_1} \tag{116}$$

In such circumstances, K_m, which equals k_{-1}/k_1 and thus K_D, provides information on the stability of the intermediate complex.

SAQ 12 (Objective 10)

(a)   If the first step is a rapidly established pre-equilibrium (recall Block 3, Section 6), then

$$\frac{k_1}{k_{-1}} = \frac{[\text{S.E}]}{[\text{E}][\text{S}]} \tag{117}$$

so

$$[\text{S.E}] = \frac{k_1[\text{E}][\text{S}]}{k_{-1}} \tag{118}$$

(b)   The concentration of the enzyme that was added at the start of the reaction, $[\text{E}]_0$, can be expressed in terms of that of the *free* enzyme, $[\text{E}]$, and that tied up in the 'nuisance complex'. (The second step has no effect on the concentration of the enzyme since the enzyme appears on *both* sides of equation 94.)

$$[\text{E}]_0 = [\text{E}] + [\text{S.E}] \tag{82}$$

Substituting for [S.E] from equation 118 gives

$$[\text{E}]_0 = [\text{E}] + \left(\frac{k_1}{k_{-1}}\right)[\text{E}][\text{S}] \tag{119}$$

$$[\text{E}]_0 = [\text{E}]\left\{1 + \frac{k_1}{k_{-1}}[\text{S}]\right\} \tag{120}$$

We can rearrange this to obtain an expression for [E]:

$$[\text{E}] = \frac{[\text{E}]_0}{\left\{1 + \dfrac{k_1}{k_{-1}}[\text{S}]\right\}} \tag{121}$$

(c)   The substrate is converted into product in the second step. Thus, the chemical rate equation for J_0, the initial rate of conversion of S into P, can be obtained from the stoichiometry of this step:

$$J_0 = k_2[\text{E}][\text{S}]_0 \tag{122}$$

Substituting for [E] from equation 121, and rearranging the resulting equation into a form similar to equation 56, by multiplying the top and bottom by (k_{-1}/k_1), gives

$$J_0 = \frac{\left(\dfrac{k_2 k_{-1}}{k_1}\right)[\text{E}]_0[\text{S}]_0}{\left\{\left(\dfrac{k_{-1}}{k_1}\right) + [\text{S}]_0\right\}} \tag{123}$$

Direct comparison of this expression with equation 56 reveals that

$$V = \left(\frac{k_{-1}}{k_1}\right)k_2[\text{E}]_0 \tag{124}$$

and

$$K_\text{m} = \frac{k_{-1}}{k_1} \tag{125}$$

Thus, the Michaelis constant K_m, provides information on the stability of the 'nuisance complex'.

ACKNOWLEDGEMENTS

Grateful acknowledgement is made to Lawrence Berkeley Laboratory/Science Photo Library for the cover photograph and to Dr David Roberts, The Open University, for the computer-generated Figure 15.